Gorillas in
Our Midst

Gorillas in Our Midst

The Story of the
Columbus Zoo Gorillas

Jeff Lyttle

Ohio State University Press
Columbus

Image on p. 25 reproduced with permission of Ringling Bros.-Barnum &
Bailey Combined Shows, Inc. RINGLING BROS. AND BARNUM & BAILEY® and
THE GREATEST SHOW ON EARTH® are federally registered trademarks and
service marks of Ringling Bros. and Barnum & Bailey Combined Shows, Inc.

Library of Congress Cataloging-in-Publication Data

Lyttle, Jeff, 1962–
    Gorillas in our midst : the story of the Columbus Zoo gorillas /
Jeff Lyttle.
      p. cm.
    Includes index.
    ISBN 0–8142–0766–9 (cloth : alk. paper).

    1. Gorilla—Ohio—Columbus. 2. Gorilla—Breeding—Ohio—
Columbus. 3. Columbus Zoological Gardens. I. Title.
QL737.P96L97 1997
599.884—dc21   97–26066
                              CIP

Text designed by John Delaine.
Type set in Adobe Sabon.
Printed by Friesens Corporation, Altona, Manitoba, Canada.

9 8 7 6 5 4 3 2 1

To Jack and Cathy
for your love, patience, and support

And to Colo,
the ultimate survivor

Contents

Foreword

The most notable lowland gorilla family in the world had its beginnings at the Columbus Zoo.

On the morning of December 22, 1956, the zoo received an early Christmas present. A baby gorilla was born, and what a momentous occasion it was! After all, this was the first gorilla born in captivity in a zoo, animal park, or any other facility throughout the world.

Colo was indeed a very special little gorilla, with large sparkling eyes. She was definitely a sensitive and frail creature to be proud of. But her birth shouldn't overshadow the accomplishments of her parents, Macombo and Millie, whose natural behaviors brought Colo into this world.

You see, back then if you were in the "zoo business" gorillas were hard to come by, and very expensive! So if you were the zoo director and lucky enough to have two gorillas, you wouldn't dare put them in the same enclosure together, especially if they had displayed aggressive behaviors previously. Remember now, the mid-fifties preceded Dian Fossey's remarkable findings about the "gentle giants" by quite a few years. So back then, conventional wisdom said that gorillas were fierce, mean beasts that would destroy anything in their paths. Recall King Kong? Placing two gorillas—even if one was a male and the other a female—in the same enclosure would certainly lead to more nasty fights where the ferocious hairy animals would tear each other limb from limb.

Fortunately, the Columbus Zoo staff had the insight to attempt another introduction between silverback Macombo and hopefully his soon-to-be darling Millie. Lo and behold, zookeepers didn't notice any fur flying, and nature took its course. Other firsts in gorilla propagation would enhance the Columbus Zoo's image over the next few decades, and soon it became "the" place for zoos to send their nonreproductive female gorillas for meetings with willing males.

I was never so astounded as when I learned that a female gorilla named Bridgette (on a breeding loan from the Henry Doorly Zoo in Omaha, Nebraska) had become pregnant by the zoo's current stud, Oscar! When an ultrasound was done, twins were revealed; another amazing first for the Columbus Zoo.

These are just a few of my memories of the precious gorillas that call the Columbus Zoo home. Throughout this book, you'll learn a lot about these magnificent animals and why we must do all we can to save them and their natural habitats. And of paramount importance to the success of the gorillas, you will understand what a critical role the zookeepers have played, for they, in a special way, are also members of the gorillas' families.

Please sit back and enjoy the true story of the Columbus Zoo gorilla family tree.

—Jack Hanna
Director Emeritus, Columbus Zoo

Acknowledgments

This book could not have been written without the time, insights, and incredible stories of the architects of the great Columbus Zoo gorilla program. More than two dozen interviews were conducted in the course of researching and writing this book, but I'd like to acknowledge the special contributions of several individuals: Beth Armstrong, Dr. James N. Baird, Dianna Frisch, Dr. Harrison Gardner, Charlene Jendry, Dr. Richard McClead, and Warren Thomas.

Thanks are also due to the Columbus Zoo for its willing participation in this project, especially Michael Pogany and the zoo's Media Productions department for providing the photographs that are vital to telling this story. All photographs that appear in this book, unless otherwise noted, are courtesy of the Columbus Zoo.

I also appreciate the contribution of the staff and research facilities at the Columbus Metropolitan Library, the Ohio Historical Society, and the Smithsonian Museum of African Art, and also the contribution of Columbus's daily newspapers, especially the *Columbus Dispatch,* which have provided extensive coverage of the Columbus Zoo and its gorillas over the past fifty years and yielded much of the anecdotal information included in this book.

Finally, I'd like to thank editors Barbara Lyons and Ruth Melville, whose commitment and style gave this book a necessary focus, and Barbara Hanrahan and Charlotte Dihoff of Ohio State University Press, who believed in this project from the beginning.

Mac

Millie

Colo

Bongo

Emmy

Oscar

Toni

Cora

The twins

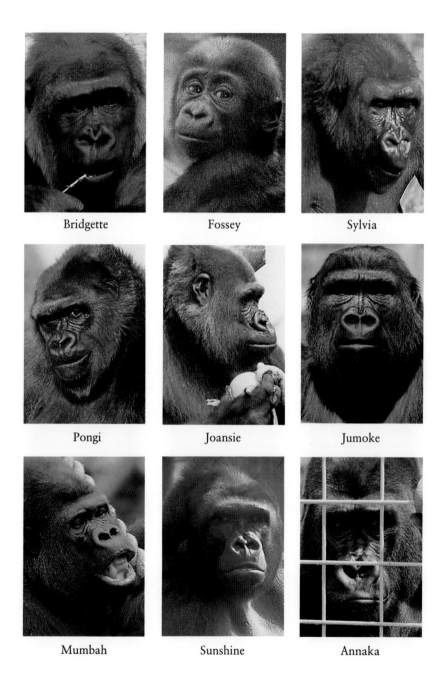

Bridgette

Fossey

Sylvia

Pongi

Joansie

Jumoke

Mumbah

Sunshine

Annaka

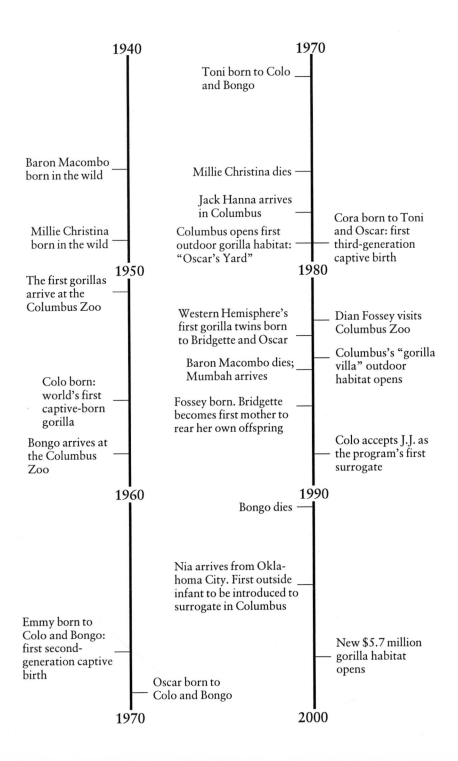

1940

Baron Macombo born in the wild —

Millie Christina born in the wild —

1950

The first gorillas arrive at the Columbus Zoo —

Colo born: world's first captive-born gorilla —

Bongo arrives at the Columbus Zoo —

1960

Emmy born to Colo and Bongo: first second-generation captive birth —

Oscar born to Colo and Bongo —

1970

1970

Toni born to Colo and Bongo —

Millie Christina dies —

Jack Hanna arrives in Columbus —

Columbus opens first outdoor gorilla habitat: "Oscar's Yard" —

1980

Western Hemisphere's first gorilla twins born to Bridgette and Oscar —

Baron Macombo dies; Mumbah arrives —

Fossey born. Bridgette becomes first mother to rear her own offspring —

1990

Bongo dies —

Nia arrives from Oklahoma City. First outside infant to be introduced to surrogate in Columbus —

2000

Cora born to Toni and Oscar: first third-generation captive birth —

Dian Fossey visits Columbus Zoo —

Columbus's "gorilla villa" outdoor habitat opens —

Colo accepts J.J. as the program's first surrogate —

New $5.7 million gorilla habitat opens —

Introduction

As we approach the twenty-first century, it has been more than a hundred years since the first gorillas were kept and displayed in captivity. In these hundred years, humans have learned a great deal about this mysterious animal. For instance, like humans, gorillas have unique fingerprints. Under a microscope, a gorilla's blood is virtually indistinguishable from that of a human. Their eyes can be hauntingly expressive and their intelligence as intimidating as their size. However, what we don't know about the gorillas may exceed our knowledge of them. Although many gorillas have lived long lives in zoos around the world, only in the last fifteen years have they been given the opportunity to live in an environment that approximates their diversified social existence in the wild. This change from solitary, unstimulating cages to functional and naturalistic habitats where diverse groups can thrive is due to the commitment of dedicated researchers in the wild, and of keepers and veterinarians in forward-thinking zoos around the world.

Much of the ignorance and misinformation that dominated gorillas' first eighty years in captivity was the result of ignorance about, or a lack of respect for, their intelligence, both of which can be seen in many popular books and films. More than fifty movies made before 1950 featured the gorilla as a villain: a fierce and horrible monster, capable of acts of superhuman violence. This cinematic image of terror—exemplified by King Kong, perhaps the ultimate and most famous gorilla exaggeration—was accepted as fact, creating an image of the gorilla sadly removed from reality. Hunters often capitalized on such myths and justified the murder of gorillas by returning from the mountains and lowlands of Africa with tales of terror and murderous attacks.

The first gorillas seen outside of Africa rarely lived more than a few weeks in captivity. Until this century, it was rare for gorillas even to survive capture and transport to Europe. The first live gorilla to be brought to the United States arrived in Boston in 1897. Barely clinging to life when it arrived, the gorilla survived in a Boston zoo for just five days. Fourteen years later, a second gorilla was brought to the United States, bound for the New York City Zoo. The animal lived only twelve days after its arrival in New York. As late as 1915, one animal expert predicted that an adult gorilla would never be seen alive in captivity, since the only animals that hunters were physically able to capture and transport from the wild were infants. It seemed improbable that such young animals could ever be cared for well enough, and long enough, to reach adulthood.

Ironically, the gorillas that lived the longest in captivity in the early twentieth century were cared for by individuals and treated as pets. John Daniel, a four-year-old male, was purchased in London by the Ringling Brothers Circus and put on display in 1921. At the time of his purchase, John Daniel weighed nearly two hundred pounds and stood almost four and a half feet tall. He had lived longer than any captive gorilla thus far. However, just one month after his purchase from a private owner, John Daniel died. His keepers at the circus, probably out of ignorance rather than cruelty, had fed the vegetarian gorilla a meat-filled human diet—including at least three drinks of whiskey per day, in an effort to keep him calm.

In the short time he was displayed at the circus, John Daniel proved amazingly popular. Many years later, when the circus was struggling to survive, John Ringling North remembered John Daniel's drawing power and began looking for another gorilla to display. In 1937 he discovered a nearly full grown gorilla that was being raised by a private citizen outside New York City. Gertrude Lintz had raised the animal like a pet, housing him in a makeshift cage in her barn. Despite his massive size, it wasn't until the gorilla, called Buddy, made a late-night escape and found his way to her bedroom that she realized she couldn't keep him any longer. She called the circus, and Ringling North purchased the eight-year-old gorilla for $10,000. Ringling North changed the gorilla's name to Gargantua the Great and proceeded to display him for the next twelve years. Capitalizing on the gorilla's big-screen image as a monster (*King Kong* had opened on Broadway in 1932), the circus promoted its new-found attraction as "The World's Most Terrifying Living Creature."

On the ship bringing him from Africa Gargantua had been scarred when a sailor threw acid in his face. The acid caused him to have a permanently upturned lip, which looked like a fierce scowl, a deformity that suited the circus's promotional

Circus poster of Gargantua, 1938. Courtesy of the Circus
World Museum-Library and Research Center.

efforts. Over the years, forty million people came to see "the largest and fiercest gorilla ever brought before the eyes of civilized man"—another of Ringling's favorite descriptions of Gargantua. Gargantua died in 1949, at the age of twenty, but not before making his mark as one of the most famous animals of the twentieth century. His strength and tenacity were legendary among circus workers, who often played catch with Gargantua until, apparently bored, he would return an underhand toss with a fierce overhand fast ball that would send his playmates ducking for cover. Gargantua could also defeat more than a dozen men in a game of tug-of-war.

Shortly after Gargantua's death, Dr. Bernard Grzimek, the director of the Frankfurt Zoo in Germany, undertook a census of gorillas living in zoological gardens. The census showed slow but steady progress being made in the care and the longevity of gorillas in captivity. But although fifty-six gorillas were living in zoos at that time, the animals were little more than curiosities to be gawked at from beyond the steel bars of a cage.

Perhaps the most famous of the era's captive gorillas was Bushman, a long-time resident of Chicago's Lincoln Park Zoo. Bushman was purchased by the zoo in 1930, for $3,600. In a matter of months, he became the zoo's star attraction, and he remained its favorite son for decades. Chicagoans were awed by Bushman, even though he did little more than eat and sleep in his concrete and steel enclosure. His keepers feared Bushman's strength and did not even enter his cage during the last ten years of his life. When Bushman fell ill, zoo veterinarians were afraid to try to anesthetize him, fearing they might kill him in the process, so they were forced to guess on a diagnosis.

When Bushman died in 1952, his age of twenty-two was thought to be a full life span, equivalent to about seventy human years. The Philadelphia Zoo's legendary male Massa disproved that myth, living to an estimated age of fifty-four (more than fifty in captivity) before his death in 1984.

In the 1950s, once gorillas could be kept in good health in captivity, the world's largest zoos turned their attention to breeding the animals. Captive breeding of chimpanzees had become relatively common, and there had also been some captive births of orangutans, though they were still fairly rare. However, the breeding of the largest member of the great ape family remained a mystery.

The Bronx Zoo seemed the most likely site for the first captive gorilla birth. The zoo was one of the biggest and most successful in the county, and its breeding pair, Makoko and Oka, had been raised together. The two had been playfully interacting with one another for nearly ten years, and the prospect of their breeding together seemed excellent. However, tragedy struck in November of 1951 when Makoko, the ten-year-old male, drowned in the moat surrounding his play yard. The loss was devastating for the Bronx Zoo and the entire zoo community.

Nearly five years passed before another chance arose, when Millie Christina, a 260-pound, seven-year-old female at the small and little-known Columbus Zoo, and her aggressive and powerful eleven-year-old mate, 380-pound Baron Macombo, were placed together in a cage. This time, Millie and Mac would make history. On December 22, 1956, Millie gave birth to Colo, the world's first captive-born gorilla.

In the more than forty years since then, twenty-six more gorillas have been born at the Columbus Zoo. Each birth is as important as the last, although, thanks to Colo and the captive-born gorillas that have followed her, much more is understood about the breeding—as well as the care and rearing—of the rare, gentle western lowland gorilla. With proper care, gorillas are capable of surviving in captivity many years beyond what is now known to be their life expectancy in the wild. At last count, nearly 350 gorillas were being housed in North America. And through the cooperation of zoos around the world, the captive gorilla gene pool promises to allow future generations to continue to study and appreciate one of nature's most fascinating creatures.

Recently, gorillas have shown their gentleness and intelligence in two widely publicized interactions with human children. In 1986 the news media worldwide replayed the protective gestures of Jambo, a full-grown, twenty-five-year-old silverback, as he placed himself between a five-year-old child who had fallen into his pen at the Jersey Zoo in the Channel Islands and the inquisitive members of his gorilla group. Ten years later, Binti Jua, a female born at the Columbus Zoo in 1988, astonished onlookers and television viewers as she carried an unconscious three-year-old boy to the safety of her keepers, while her own juvenile offspring clung to her back. The boy had fallen into the outdoor gorilla habitat at Chicago's Brookfield Zoo, where Binti was on breeding loan. Binti's actions moved hundreds of people to write to her and the zoo, praising her actions and offering donations. An eighty-eight-year-old woman from Los Angeles wrote: "My dearly loved Binti. I had to cry with love for your beautiful gesture." Despite living on social security, the woman enclosed $20, asking Binti's keepers to buy the gorilla a "special dish of ice cream."

Just as Hollywood hype inaccurately led to the gorillas' reputation for viciousness, the modern media hype over Binti's protection of an injured child may also have done gorillas a disservice, albeit more subtly and unintentionally. Anthropomorphic coverage of Binti attributed her actions to maternal instincts thought to result from her human rearing in the Columbus Zoo nursery and her subsequent interaction with humans at other zoos. However, Binti's keepers in Columbus, San Francisco (where she lived for several years), and Chicago all downplayed their influence on her actions. While it's often easy to ascribe human traits to one of our closest primate relatives, most keepers make it clear that this tendency—even when intended as a compliment—is generally misguided. Most acknowledge the fine line they walk with all captive animals, especially primates, in an effort to care for them properly without treating them like humans. They must fight the urge to project their own feelings of fear, isolation, joy, remorse, and even love onto the animals. At the same time, though, keepers must rely on those same emotions when determining the best possible living environment for the gorillas—as gorillas.

Throughout this book there are references to human influences on gorillas, both positive and negative, and to human interpretations of their behaviors. Gorillas have benefited from the practices of human medical science, for example, and have also suffered the indignity of being displayed wearing human clothing. The controversy over ascribing human characteristics to any animal—even domestic dogs and cats—goes to the heart of the argument over whether animals should even be kept in captivity. Many people believe that no matter how zoos treat animals, zoos are nothing more than self-serving human endeavors. It's an argument that will exist as long as zoos exist, and one that is recognized by the keepers who care for the animals every day. "The interaction between people and gorillas is important only because of what it means to the gorillas," says long-time Columbus keeper and educator Charlene Jendry. "If the interaction is only for the human's benefit, it is wrong. You've stepped over the line."

Whether you believe in zoos or not, the fact remains that modern zoos have important responsibilities as conservators and educators; their days of functioning simply as animal menageries are gone. An examination of the evolution of the Columbus gorilla breeding and conservancy program is also an examination of the evolution of zoo practices in general. The Columbus Zoo has embraced and nurtured its fame as a center for gorilla breeding while sharing its accumulated knowledge with the rest of the zoological community. The world's gorilla population, and those of us who appreciate the opportunity to observe and learn about one of nature's most intriguing animals, owe a great debt to the Columbus Zoo gorilla family and their keepers.

1

Gorilla Bill

In the fall of 1950, William Presley Said was hungry and broke. Worse, he feared he was losing his sanity. He had spent nearly six months hunting in the dense forests of West Africa, surviving on the hope that he would find and capture the powerful, mysterious gorilla. If he could capture gorillas and get them back to the United States, the hunt would make him rich. Said was in the dangerous business of gorilla hunting for the money, pure and simple. However, after an agonizing summer and fall, Said had caught only half a dozen chimpanzees, all of whom eventually escaped, plus a nearly fatal case of malaria.

Bill Said had always equated hunting with financial opportunity. At age six, he had owned a .22 bolt-action children's rifle. By the the time he was ten, he was killing pheasants and rabbits with a single-shot 12 gauge. While his friends in Columbus, Ohio, were sweating it out making a few dimes delivering the afternoon newspaper, Said was setting up efficient traps to catch muskrat, fox, raccoons, and skunks. Before he had finished grade school, he had made $200 selling muskrat

11

pelts. Said always chuckled when he told that story, saying his friends would have had to work all their lives selling newspapers to make $200.

In 1947, after receiving a medical discharge from the army owing to chronic asthma and attending two colleges in an unsuccessful attempt to earn a degree in education, Said took a civilian job at an American air base in Frankfurt, Germany. Frankfurt was Said's fourth new city in five years. The only constant in his wayward existence was his love for hunting. Frankfurt not only offered Said the opportunity for steady work but also gave him access to a whole new area of the world to hunt in.

In Germany Said found a group of friends who shared his passion for hunting. Together they enjoyed several successful hunts in central and southern Germany and in the north of France. But even during his most successful hunting trips, Said kept talking about what a thrill it would be to hunt for the really big game in the wilds of Africa. Part of the lure of working in Germany was the fact that Europe was one step closer to Africa.

Said rallied his buddies, and together they planned an African hunting expedition in 1948. Even when his friends all backed out at the last minute, Said was determined to go on alone. Loaded down with guns and camping equipment, he boarded a train for Munich, then another for Rome. From Rome, he took a plane to Cairo. Though he had made no arrangements for a safari, he was excited to have finally made it to Africa.

As luck would have it, one of the employees at Said's hotel in Cairo knew a few things about African hunting. He had grown up in a region some 1,500 miles to south along the Nile

River in Anglo-Egyptian Sudan. The man, whose deep scars on his cheek were ritual markings that had been made with a leopard's claw, told Said that big game was plentiful around Juba. It was just the environment Said was looking for. The next day he boarded a plane for Khartoum, then took a bus to Juba. In Juba, Said hooked up with a member of the African game department who often led American GIs on hunting trips. The pair spent two months along the Nile, killing several antelopes and three elephants. Said also managed to make a live capture of a female leopard and her two cubs.

The leopard mother died during her transport back to Europe, but Said was able to sell the cubs to the Frankfurt Zoo for $300 apiece. He later told a Columbus newspaper reporter that their capture was surprisingly easy—and surprisingly lucrative. "If I had known [how much they were worth]," he said, "I would have brought back 20 leopard cubs." Said also made $1,200 selling the ivory of the elephants he had killed. His total take of $1,800 was enough to finance his trip, and it convinced him that a good living could be made as an African big-game hunter.

At the age of twenty-three, Said decided his experience with the African game commissioner had taught him enough about Africa to lead hunting expeditions himself. He opened a big-game hunting business in Leopoldville, the city known today as Kinshasa, in the present-day Democratic Republic of the Congo (formerly Zaire). The skinny, fast-talking Catholic boy from the American Midwest spent the next two years as a guide, earning a reputation as a reckless and adventuresome hunter. As a white American, Said stood out among the locals and other African hunters. His unique looks and renegade reputation made him a well-known figure in the area.

In early 1950, while sitting around a camp fire midway through an expedition, a Frenchman told Said that if he really wanted to make money as a hunter, he should go after the biggest prize in the forest—the gorilla. Only a handful of zoos in Europe were lucky enough to have gorillas in their animal collections, and dozens of American zoos were desperate to add the immensely popular animal to their menageries. Most zoos, the man said, would pay $5,000 or more per gorilla. Suddenly the money Said had earned capturing leopards and killing elephants seemed small compared to the promised riches that could be reaped from gorilla hunting. Said decided to find out more. He closed his business and moved back to Columbus to contemplate his gorilla hunting strategy.

The Frenchman had told Said that Gabon was the most gorilla-rich country in West Africa, but Said soon discovered that the French government required a sound scientific reason from a reputable organization to permit the taking of gorillas from Gabon. He went to work, contacting several universities and selling his skills as an accomplished hunter and businessman, hoping a university would "sponsor" his first hunt. In fairly short order, Said secured a commitment from the University of Wisconsin. The university agreed to pay $2,000 up front for expedition expenses, and $2,000 per live gorilla upon receipt of the animals. According to Said's permit application, the university wanted the animals for scientific and psychological experimentation. It wasn't the $5,000 per animal Said had hoped for, but he was guaranteed a decent return if he had a successful hunt.

In the early summer of 1950 Said got back on a plane for Africa. When he arrived in Leopoldville he purchased an old, run-down ambulance for use as a transport vehicle for the

Bill Said, 1950. Credit: *Columbus Dispatch*

captured animals. He also hired a cook, an interpreter, and several local men to serve as guides and helpers. Two days after his arrival, Said headed for the dense forests of Gabon in search of the elusive gorilla.

Said's hunting party searched several weeks without even sighting a gorilla. They had seen plenty of signs of the animal, including a banana tree that a gorilla had apparently ripped apart to get to its pulpy interior. It took eight of Said's men to do the same thing to a tree of equal size. If they weren't in awe of the gorilla's strength already, the mangled banana tree convinced them that they were up against one of nature's strongest creatures.

While searching for gorillas, Said's group had managed to capture six chimpanzees, in hopes of selling them for some extra cash and prolonging the gorilla hunt. But before they could be sold, all six managed to elude their captors in one mass escape from the converted ambulance. Said's first gorilla hunt was turning out to be a bust.

On a hot September night, a frustrated and fever-ridden Said went to sleep at his forest camp. It was two and a half weeks before he awoke again, lying in a hospital bed eighty miles away. As Said had clung to life in a malaria-induced coma, members of his expedition crew had put him on a stretcher and carried him from the crew's base camp in Komono four days through the African lowlands before finally reaching the nearest hospital in Sabeti, French Equatorial Africa.

Said was not well enough to leave the hospital for two more weeks. His health was back, but little else. He was broke and unsure if he'd ever have the opportunity to return to the forest. However, back in Columbus, Said's father, Kenneth, known to friends as K.C., was not ready to see his son give up on his dream. K.C. loaned Bill enough money to reoutfit his crew and restart the expedition.

Bill headed back to the forest on Friday, October 13. A week of hunting passed, followed by another and another. The crew waded through several chest-deep and snake-infested streams, battled the stings of thousands of black flies, and warded off ants that could kill a chicken in minutes. They slept in makeshift beds outside of mud-plastered huts, it being too hot to sleep inside the huts. The living conditions were wearing Said down, and so was the fact that his group had still not spied a gorilla. As the pressure mounted, he considered giving up the hunt and going back to Ohio for good.

In a last-ditch effort to salvage the hunt, Said decided to engage the services of a local group of people known as the Bacola. They were excellent hunters and fishermen who knew the dense lowlands of the Congo region better than anyone. Although the Bacola feared the gorilla and had never hunted with a white man, Said assured them that he would physically capture the gorillas if they would help him find them. To confirm Said's commitment to the Bacola and to the hunt, the Bacola chief asked for the "blood of the white gorilla hunter." When Said agreed, the chief used a large knife to make a ten-inch incision in Said's arm. Said was scarred for life, but for the first time in months he was optimistic that he would find gorillas.

Within days, the Bacola had located a troop of western lowland gorillas. Gorillas travel in troops of as many as twenty or twenty-five, led by a mature male known in the lowlands as the garçon. The garçon is usually the largest male, and his back is covered with brilliant silver hair, a sign of gorilla maturity. Today, these gorilla leaders are commonly known as silverbacks. The Bacola located the gorillas at night by listening for the snore-like grunt the animals make while sleeping. In the wild, the older animals sleep on the ground in fresh nests made each day. The younger animals often sleep in nearby trees.

Once the Bacola located the gorilla troop, more than a hundred men encircled the sleeping animals and staked out their places for a dawn attack. They knew that when the garçon sensed their presence, he would charge angrily. Many of the Bacola had seen an angry garçon before and knew there was no more intimidating animal in all of Africa.

As Said told the story later, the Bacola performed an elaborate pre-hunt ceremony to prepare for the battle ahead. The chief, his face covered with white paint, gathered his hunters around a large fire. One by one, the chief poured a white liquid into the eyes of his hunters. The liquid was held in a large leaf, rolled up like a funnel, which the chief held high above his head after administering each dose. Said's interpreter told him the fluid was supposed to give the hunters clear vision during the predawn darkness, at the crucial final moments before they engaged the great apes. The white fluid was also poured into the fire, and the hunters were told to study the flames and the smoke for signs of the hunt. As the ceremony continued, a small animal was sacrificed and its blood spread on the hunters. The blood was thought to give them the necessary courage and strength to battle the gorillas.

Said also watched with interest as the children of the tribe played a game of "white hunter." Using the mid-rib of the banana leaf as a make-believe gun barrel, the children threw twigs at the tribe's wild barking dogs, pretending the dogs were gorillas.

At dawn, on November 11, 1950, the hunters quietly took their places. Said and members of his expedition team were not allowed near the hunt site until the Bacola were in position and had raised their hemp net. When the net was ready, Said was summoned, and the hunters slowly tightened their circle around the sleeping animals. When the silverback became aware of their presence, he charged, as expected, in an all-out effort to protect his troop. Recollecting the silverback's attack, Said shuddered, calling him a "terrible beast." "He beats his chest and charges on all fours, screaming as he runs. Horror strikes you. I never cease to fear him," he told a Columbus newspaper reporter. "You know you are dealing with an animal that has a keen mind. I hate to kill the gorilla, but you've got to get rid of the big ones or they'll get rid of you."

Using spears, the hunters killed the silverback, creating

chaos in the troop and allowing Said to capitalize on the confusion to capture as many of the young gorillas as he could. The only way Said knew how to capture them was with his bare hands, subduing them by wrestling them to the ground.

One young male gorilla, whom Said named Macombo after the city (Mekambo) in which he was captured, was taken alive. After many months of suffering, Said had finally captured his first gorilla—but only one, not the five or six he had hoped for. Still, he had captured a healthy seventy-pound gorilla, thought to be four or five years old. The animal was placed in a cage made of sticks and vines and carried back to camp.

Two members of the hunting party had been bitten during the chaotic capture, and eight members of Macombo's gorilla family had been killed. It had been a bloody massacre, more bloody, in fact, than Said had expected. Gorilla bodies had lain strewn about in a hundred-meter circle. The hunting party had taken down trees, trying to reach frightened animals that had climbed them for safety. One animal had been killed when a falling tree had crushed it. Said had found Macombo when he heard the animal whimpering among the branches of a fallen tree.

The remains of the dead gorillas were strapped to sticks and carried back to camp, in clear view of the captured animal. Macombo cowered in his makeshift cage while the expedition team prepared and cooked the dead gorillas in a post-hunt ceremony. Said later told a reporter that the gorilla meat he ate was the "finest meat I've ever eaten anywhere."

It was the first time the Bacola had seen a gorilla captured alive. It was also the first time any member of the tribe had killed a gorilla with a spear. Though the hunt had succeeded in capturing a gorilla, and the dead gorillas would provide food for the Bacola for several days, they did not like the experience. They decided to abandon Said and his gorilla hunting. Said paid them for their help by giving them army surplus shirts he had purchased in Germany. On the back of the shirts were the letters "P.O.W."

Undeterred by the loss of the Bacola's assistance, Said and his expedition party moved on to an area near the city of Boma, about fifty miles from Leopoldville along the Congo River. There Said engaged the services of another group of local people, the Mbete. On December 4, the party captured five more animals in another bloody hunt. During the melee Said was attacked by a 240-pound female gorilla. As the natives watched, Said, though bitten in the ankle, killed the animal with a spear. The act of single-handedly killing the gorilla made him a hero in the eyes of the Mbete, and his reputation as a fearless hunter grew.

In the days after the second hunt, one of the caged gorillas died from injuries it had suffered during its capture. Another was killed by Mbete hunters after it escaped and bit one of them. Said and his hunting party had killed more than twenty animals in the process of capturing the four live gorillas now in his possession.

Satisfied that he had captured all he could on a single expedition, Said put his four gorillas—two males and two females—on an airplane headed for the United States. As the plane made a stopover in Brussels, Said checked on his animals and found one of the females dead; she had frozen to death during the flight. Now down to only three live animals, Said was further disappointed when he discovered, on arriving in New York, that the University of Wisconsin could not afford the proposed gorilla experiments and was therefore no longer interested in purchasing his gorillas. On December 22, 1950, the university officially released its rights to the animals.

The same day Said called his friend Earle Davis, the superintendent of the Columbus Zoo, to offer him a deal. He told Davis he was in New York City with three gorillas that he would sell to the Columbus Zoo for a total of $10,000. He described the circumstances of the animals' capture and gave Davis their vital statistics: one male approximately five years old and completely healthy; another male, about eighteen months old, suffering a bit from the effects of travel and nursing a small gunshot wound suffered during his capture; and a female, thought to be twenty-one months old, also a little sick from the stress of capture and travel. Said claimed that he could easily get $15,000 for the animals if he sold them individually to other zoos, but since Columbus was Said's hometown zoo, and Davis was his friend, he'd cut him a deal. As part of the deal, Said also wanted to spend the next month at the zoo, observing the animals and learning all he could about their habits before heading out on his next hunt.

Davis was delighted. As manager of the city's zoo, he was known as a dedicated public servant. His keepers considered him impulsive and excitable, a "fly by the seat of your pants" kind of guy. Knowing what the addition of three gorillas could mean to his zoo, Davis struck the deal over the telephone. Ten thousand dollars was quite a large investment for Davis's small zoo to make, but he felt strongly that the animals would be more than worth the investment.

In a matter of minutes, the Columbus Zoo went from having no gorillas to having one of the largest gorilla collections in the world. Most American zoos that had gorillas, even the big ones in New York, Chicago, Philadelphia, and St. Louis, only had one or two. Suddenly, Columbus would have three, the fifty-eight, fifty-ninth, and sixtieth gorillas to be brought to the United States. While it was mostly a case of blind luck, many zoo experts argue that three turned out to be a magic number that led to the successful captive breeding of gorillas. Since so many zoos had tried unsuccessfully for years to breed their gorilla pairs, experts have theorized that when there are just two animals, the pair forms a sister-and-brother relationship; but when three or more gorillas are brought together, the animals begin pairing up and mating. Today, keepers say three is the minimum number they like to see in troops (although the best captive groups have a dozen or more members, of appropriate ages and social status).

Said told Davis that he would bring the healthy Macombo to Columbus with him but that Davis would have to help get the other two animals to the zoo after they had had a week or so to recover from their journey. Davis readily agreed, since he was already scheduled to be in New York on a zoo buying trip the week after Christmas.

On January 4, 1951, Columbus's newspapers ran front-page stories welcoming the zoo's first gorilla to the city. Macombo was described as a five year old with a voracious appetite who spent his first day in Columbus eating fresh fruit—bananas, oranges, grapefruit, and apples—and brown bread. Said had transported him via a heated rail car and a heated truck. On arrival, Said warned that the animal's size belied his incredible strength. Though he didn't look much larger than a full-grown chimpanzee, Mac had thrown one of the members of Said's hunting expedition high in the air during a foiled escape attempt.

Nine days after Mac's arrival, more front-page stories hailed the arrival of the other two gorillas: a female named Boma, for the city nearest her capture, and a male now called Buckeye, in honor of the state tree of Ohio and the mascot of The Ohio State University. Earle Davis had brought the two animals to Columbus in orange crates, which he had secretly carried into his Pullman car as luggage. He did his best to hide the contents of his crates by wrapping them up like gifts. Davis said the animals let out an occasional "whoop," which puzzled the conductor and a few passengers, but the gorillas remained undiscovered.

Upon arriving at Columbus's Union Station, Davis hailed a cab for the twenty-five-mile ride from downtown Columbus to the zoo, located along the Scioto River in Powell, a small community northwest of the city. Despite the investment he'd made in the gorillas, Davis hadn't made any special arrangements for transporting them from the train station to the zoo. "Every time we stopped for a traffic light, they wailed," Davis told one of the local newspapers.

Neither Boma nor Buckeye wanted to get out of their crates when they finally arrived at the zoo. Buckeye poked his head out of the box, then ducked back in, whimpering like a frightened child—understandable behavior for an animal that had been stolen from its family and freedom in Africa and placed in dark wooden boxes for nearly a month of captivity and travel.

Davis announced to the citizens of Columbus that they would be able to see their zoo's three gorillas when the park opened in March. Meanwhile, he asked Columbus residents to help the zoo give the animals official names in a "Name the Gorillas" contest. A committee of city officials, Davis among them, took three weeks to sift through thousands of entries

Earle Davis and Buckeye, who arrived in Columbus in
an orange crate

before coming up with three winners. Macombo would be
known as Baron Macombo, or "The Baron." The winning
name was submitted by a nine-year-old boy. The zoo's other
male gorilla, formerly known as Buckeye, would be given the
name Christopher, in honor of the city's namesake, Christopher Columbus. Three young girls were given credit for the
name. The committee chose the name Christina for the zoo's
female gorilla. Her name, the female counterpart to the name
Christopher, was submitted by a seven-year-old boy. The

winners of the "Name the Gorilla" contest were given gold-plated zoo memberships, good for free admission to the zoo for life.

The animals were put in the zoo's former carnivore building, known as the Pine Building. More than twenty-five years old, it was one of the zoo's original structures, and the cages didn't have drains. One keeper remembers how difficult it was to clean the gorilla's quarters. The routine required one keeper to engage the animals' attention while another hosed the area down. The keeper who kept Mac busy usually ended up in a "playful" wrestling match. Luckily, Mac had lost his deciduous teeth and had not yet grown permanent ones. Still, the keepers who had the job of attending to Mac would leave the cage with large red gum marks up and down their arms.

Christopher and Millie Christina

It didn't take long for Baron Macombo, who weighed more than a hundred pounds a few months after his arrival in Columbus, to give Davis a firsthand example of his strength. He began mangling his Pine Building cage and its ⅝-inch steel bars. In addition to twisting the bars, Mac managed to pull them from their concrete moorings. Searching for reinforcement, the zoo superintendent checked with the Ohio State Penitentiary, located in downtown Columbus, for advice on building a more sturdy cage that would hold the Baron. However, Davis said he wasn't sure that even Ohio's most hardened criminals could match the strength of Baron Macombo.

The keepers forgave the gorillas' destructive behavior, largely because they attributed their violence to boredom. Although the animals could hear one another, since the cages were side by side, concrete walls prevented them from seeing one another. The keepers say that the three loved to perform for people and only tore up their cages during the winter months, when the zoo was closed. The keepers feared that the gorillas were so strong that they might eventually free themselves by brute force.

In 1953 nearly half a million people visited the Columbus Zoo, and the gorillas were the star attractions. On the strength of the gorillas' popularity, the zoo ran a membership drive that raised the total membership to 43,000. The fee for an annual membership in 1953 was $1.

After three years of trying to fix and update the outdated Pine Building, zoo officials decided that the growing animals would have to be moved to a more modern building. By April of 1954, the zoo had prepared cages for the gorillas in the new Arthur C. Johnson Aquarium. Ten feet by fifteen feet in size, they were equipped with ¾-inch—and hopefully Baron-

Baron Macombo, whose distinctive looks can be seen in five generations of Columbus gorillas

proof—steel bars. Though small, the new cages would be sturdier and easier to clean.

When the gorillas' new home was ready to be occupied, Davis and his keepers faced the difficult challenge of moving the three animals across the grounds. Baron Macombo, for one, now weighed more than two hundred pounds. Davis decided the easiest way to get the animals across the park was to construct a wooden crate, complete with a trap door. He ordered that the gorillas be coaxed, one by one, inside the crate. Once an animal was inside, it was to be moved to its new cage.

Davis watched as the keepers carefully placed the crate at the opening of Christopher's cage and placed food inside it. As keepers looked on, the gorilla moved inside the box toward the food as the trap door closed. In a matter of minutes,

Mac sitting inside his cage in Columbus's first great apes
building

he was ready for transport. Christina, who had taken on the
nickname Millie in honor of Earle Davis's gorilla-loving wife,
Mildred, was next to go. She also went peacefully, taking the
bait as the trap door closed behind her, and the crate was
quickly moved across the park to the revamped aquarium.

Baron Macombo, on the other hand, would not be as easy
to transport. He had watched the keepers capture the other
two and seemed to understand how the system worked.
When the keepers put food in the box to coax Mac inside, he
went after it, just like the others, but as he took the bait he

kept one leg strategically extended behind him, preventing the trap door from falling. Davis, amused at first, next tried to lure Mac into the box by using a playful chimpanzee. The chimp aroused Mac's curiosity, but not enough to get him all the way into the box. Finally Davis decided that the zoo's female gorilla would be the ultimate bait. However, having fallen for the trap door trick once, Millie Christina could not be fooled a second time, and the keepers couldn't get her out of her *new* quarters.

In a last-ditch effort, Davis went to the newspapers and asked the public for their ideas. "Anybody know how to move a gorilla?" he asked.

Baron Macombo continued to outwit his human keepers for several days. Eventually, it took the use of a large snake to scare the Baron into seeking refuge in the crate. Once the door closed him inside, he sat calmly for the duration of his short move.

Soon after the animals were in place in their new quarters, Davis decided that three growing gorillas were too many for the zoo to handle. He engineered a trade of Christopher to the zoo in Basel, Switzerland, in exchange for two rhinoceroses and two cheetahs. One of the rhinos, named Clyde, is still a part of Columbus's rhino collection and the oldest animal at the zoo. Christopher, renamed Stefi, would sire Goma, the world's second captive-born gorilla. Stefi sired four additional offspring at the Basel Zoo before his death in 1981 at the age of thirty-two.

Meanwhile, the man who captured the three gorillas had become somewhat of a celebrity. Now known as "Gorilla Bill," Bill Said was featured in a five-page spread in *Life* magazine in 1951 entitled "*Life* Goes on a Gorilla Hunt." He had also appeared on the network television programs *What's My Line?* and *We the People*. In four separate expeditions, including his original hunt, Said had captured eighteen gorillas. In addition to Columbus, zoos in Chicago, Cincinnati, and Toronto, as well as Yale University and the Brazzaville Zoo in Africa, had bought animals from Said.

Like many gorilla hunters of his era, Said would often fabricate a scientific pretext for his hunts by using the real names of university professors on his hunting applications. Once the animals were captured and transported to the United States, he would claim that a difference of opinion had ended the original agreement. At that point, he would be free to sell the animals to the highest bidder. In one case, Said used the name of Ohio State University professor Morgan Allison on an application to the French government, indicating that Allison and Ohio State wanted the gorillas for psychological testing. Allison was in fact a professor in the College of Dentistry. Allison's wife says the university was not very happy when they found out what Said had done, but by that time Said had used his ill-gotten hunting permit to capture and sell three gorillas.

Just before Christmas of 1951, Said returned to Columbus to recover from a case of gangrene suffered as the result of a gorilla bite. He brought two infant gorillas with him, which were to stay at his parents' home. A third gorilla was already there, whom Said's mother, Ethel, had been nursing back to health after the ordeal of its capture the previous spring. Said thought the infant was the youngest gorilla in captivity, estimating that it was just a few days old when he captured it. By this time Said's father had quit his regular job to act as Bill's animal broker, matching captured animals with interested zoos.

Back in Columbus, Said boasted to a local newspaper that he had grossed $50,000 hunting gorillas in 1951, and that he would be "on easy street if four others hadn't died before we could sell them." That was a princely sum in 1951, comparable to the salary of the greatest baseball player of the era, Joe DiMaggio. In the course of his four hunts, Said had been through a lot. "I think my luck has just about run out," he told the reporter. "The gorilla country is the worst I've ever been in. You get fed up mentally and nearly lose your mind. I lost 40 pounds during the last safari." He planned to take one more safari and then call it quits. He said hunting was too dangerous and had resulted in too many injuries within his hunting party. The gorilla hunter said he wanted to set up his own zoo in the South or Southwest and leave the wilds of Africa to other hunters who hadn't used up all their luck. "I've had my fill of big-game hunting. I think it's time I settle down while I'm still in one piece," he said.

Said's words were tragically prophetic. He was killed in a truck accident while acting as a consultant on the John Ford film *Mogambo,* in April 1952. Said, only twenty-six years old, was reportedly on his way to the airport to pick up the film's actors, who included Clark Gable, Ava Gardner, and Grace Kelly. After receiving a wire from Said's assistant informing them of the accident that had taken Bill's life, his parents made arrangements to have his body flown back to Columbus for burial. Two of the gorillas Bill had captured were still living in the Said home when their son's body was laid to rest.

Said's friends in Columbus feared that his death may not have been an accident. The gorilla hunting business was brutal, and Said had managed to make a good deal of money exploiting the fertile African lowlands and the hunting skills of its people. Certainly he had made enemies. But local officials and the American Embassy in Leopoldville could find no evidence that the truck accident that killed Bill Said was anything but an accident.

Although it is true that the circumstances of Mac's and Millie's capture were horrible, their presence at the Columbus Zoo—and the breeding program they helped to start—have done much to teach the world about the magnificent gorilla. After surviving the rigors of their capture and transport to Columbus, the two gorillas adapted well enough to their new environment to make history.

2

Colo: The Celebrity Newborn

Things weren't going so well for Earle Davis in 1956. Attendance at the zoo was high, and his zoo collection was growing, but the annual report from the zoo's veterinarian, Robert Vesper, submitted to the Columbus Zoo Commission, was devastating. Among other things, Vesper reported that the zoo was marred by "bad housing, poor sanitation and malnutrition." He blamed the poor conditions on "too many animals and not enough help." Furthermore, Vesper, who was on retainer and made only weekly visits to the zoo unless there was an emergency, contended that animal deaths were going unreported. Although thirty-seven animal deaths had been reported over the previous thirteen months, the veterinarian wrote that he was sure there had been others that Davis had not told him about. Further, Vesper demanded that postmortem exams be conducted when zoo animals died, arguing that knowing the reasons for animals' deaths could help to prevent further illness and death.

33

The report caused quite a furor at Columbus City Hall, and Davis was the target of much of the anger. Sid Phillips, the head of the commission, told a local newspaper, "The commission only wants the zoo cleaned up, that's all. If your house is dirty, whom do you blame?"

In Davis's defense, Vesper said the zoo was understaffed. Davis agreed, adding that he had ordered his staff to take care of the animals first and the grounds and buildings afterward. "You'll wade in popcorn up to your knees, but the animals will come first," he said. Meanwhile, an event was unfolding that would turn the year around for Davis and the zoo, and make people forget the controversy surrounding the veterinarian's report.

What Davis didn't know at the time was that one of his part-time keepers, a twenty-five-year-old veterinary student named Warren Thomas, had a secret. And with each passing day, his secret was getting bigger. Without telling Davis or anyone else at the zoo, Thomas had allowed Millie Christina, then eight years old, to get pregnant.

Millie and Mac, who was now eleven years old, weren't supposed to be together under any circumstances. After a couple of failed mating experiments, Davis had strictly forbidden keepers to allow the two to share a cage. On the two occasions that Davis had put the gorillas together, the much larger Mac had acted very aggressively toward Millie. Both encounters were frightening for Davis and everyone who witnessed them. (Today, gorilla keepers know that gorillas are capable of very rough play. It may be that what Davis interpreted as dangerous aggression was little more than normal roughhousing between two strong animals.)

Gorilla breeding was the goal of every zoo with a male-female pair of gorillas, but no zoo was sure how to properly facilitate the act. Nobody knew what conditions enabled these animals to breed, and there had never been a captive birth—not even a captive pregnancy. All zoos knew how to do was to keep the animals healthy, put them together when the female seemed to be ovulating, and hope. In Columbus, Davis's insistence that his two gorillas be separated made breeding impossible. He made it clear that he was not willing to risk losing one or both of his prized animals on the off chance that a miracle would occur.

However, in the months following Davis's edict, Thomas had observed, at about the same time each month, several playful encounters between the two animals. Millie would get excited and back her rump up against the door of Mac's cage, and Mac would gently play with her. Thomas didn't know if she was ovulating or not, but it looked like a mating ritual to him. Thomas tried to tell Davis about the gentle interactions he had observed, but Davis wouldn't hear of putting the two animals together again.

Thomas simply refused to take no for an answer. "I was a brash young kid. The arrogance of youth clouded my thinking," Thomas says forty years later. "When I knew the time was right, another keeper and I quietly started slipping them together, either early in the morning, or at night, before we went home." Imagine the blind confidence of a young man willing to leave the zoo's two prized animals together for an entire night! When he arrived at the zoo the next morning, he simply put them back in their own cages, and no one was the wiser.

Thomas, who later became a director at several zoos—including Los Angeles, where he spent thirteen years as director—now admits he took an incredible chance in disobeying Davis's directives. Looking back, he says he would have severely disciplined any employee of his own who'd have dared pull such a stunt, putting two incredibly valuable animals at risk on the basis of a hunch.

However, the stunt had apparently worked. After two or three secret rendezvous, Thomas noticed that Millie's monthly "mating rituals" stopped, and he began to suspect that this might mean she was pregnant. Had the unthinkable actually happened? Thomas knew the prospect would make him both a hero and reveal him as insubordinate.

He spent every spare moment for the next seven and a half months watching Millie and trying to muster the courage to tell Davis what he had done. Thomas finally decided that if his theory was true and Millie was pregnant, he had to tell somebody. In a matter of weeks, the Columbus Zoo was going to have the first gorilla ever born in captivity. "I was excited about the prospect of the captive birth of a gorilla," Thomas says. "But I knew I had to tell Mr. Davis before it happened. And I knew he was smart enough to figure out that it wasn't the result of an immaculate conception. He would certainly figure out that it was me who had defied his orders and let the gorillas mate."

When Thomas finally told the zoo director that Millie was pregnant, Davis exploded—with joy. The director wanted to tell everybody, and despite the risks that Thomas had taken, he never said a cross word to his bold part-time keeper. Thomas says Davis never even asked him to explain how it happened. He was just too elated to care.

As the zoo edged closer to history's first captive gorilla birth, it was dogged by many unanswered questions. Most important, no one knew the gestation period of a gorilla. The world's limited knowledge of the animals in the wild, combined with the fact that no gorilla had ever been pregnant in a zoo, forced Davis and his staff simply to guess. Thomas told Davis he was pretty sure that the date of conception was April 8, so on the basis of the human gestation period of nine months, Millie's due date was set. Davis announced that on or about January 8, 1957, the Columbus Zoo was expecting a new baby gorilla.

The zoological community was outwardly excited but privately skeptical. The big zoos in places like Washington, New York, and Philadelphia were sure they would be the first to have a gorilla conceive in captivity. How could a small zoo like Columbus have figured out the secret? Zoo directors from all over the world began calling to find out what Columbus had done to enable its gorillas to breed successfully.

On the last Saturday before Christmas 1956, shortly before eight o'clock, Thomas was making his usual morning feeding rounds. Even though it was officially just the first day of winter, Columbus was in the middle of a holiday cold spell, and it was particularly cold that morning. Thomas had agreed to meet his friend and fellow keeper Terry Strawser at the reptile house for a cup of coffee after he fed the gorillas. When Thomas finished giving Mac and Millie their breakfast of mixed vegetables, he topped it off with a hard-boiled egg. As he rubbed his bare hands together to warm them, Thomas wondered why Millie hadn't shown much interest in her breakfast. She was usually an enthusiastic eater. Although he didn't have a clear view of her from the back of the cage, she also seemed to be moving about her cage rather clumsily and acting strangely. As he was about to leave for his coffee date, Thomas decided to walk around to the front of the animals' cages to get a better look.

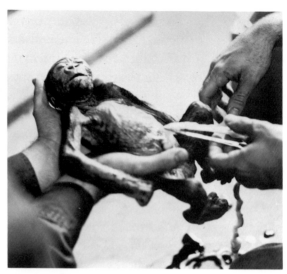

Cutting Colo's umbilical cord

Mac looked fine in his cage, but Millie had a glassy look in her eye. She seemed distant and frightened. Then Thomas looked down. There, lying on the cage's cold cement floor, was an amniotic sac with a baby gorilla still inside it. Millie had given birth while an oblivious Thomas was inside the gorilla house.

When Thomas discovered the infant, its umbilical cord was trailing away from the sac it was lying in. "Holy God," he thought to himself, "the first gorilla baby ever born in captivity, and it was born dead." Thomas's heart sank. As he stared at that sac and contemplated how close the zoo had come to making history, the sac moved. "That baby's alive!" he shouted, to nobody and everybody.

The baby couldn't have been more than a few minutes old. It was simply pure luck that anyone was in the gorilla house when Millie gave birth. During the winter months, the animals were usually visited only during feeding time and when their cages were cleaned. If Millie had given birth in the night, it is unlikely her baby would have survived until morning. Today, twenty-four-hour watches are instituted well before a gorilla is due to give birth, but in 1956 there was no precedent for planning a gorilla birth.

Thomas hurriedly lured Millie to her rest cage, away from the area where the baby was lying, and slammed the door. Thomas would find out later that he failed to lock Millie in her cage, but she was probably in postpartum shock, barely recognizing what was going on around her. Thomas was nearly in shock as well. His mind and heart were racing as he scooped up the baby, sac and all, and rushed it to the kitchen

Millie, hours after giving birth to Colo

area of the gorilla house. As quickly as he could, he broke through the sac and sponged the infant off. It was a female.

As the newborn struggled, breathing only sporadically, Thomas gave her a vigorous massage. "I could feel her life ebbing in my hands," he says. "She wasn't doing very well breathing on her own. I had to get her lungs inflated with oxygen, and the only way I knew how to do it was with mouth-to-mouth resuscitation." Thomas was so excited and filled with emotion, he nearly blew the three-and-a-half pound gorilla up like a balloon. His emotional bursts of air caused her to hyperventilate.

Meanwhile, Strawser, wondering why his friend hadn't shown up for coffee, wandered over to the gorilla house and peered through a window at the ape house entrance. Out of the corner of his eye, Thomas could see his friend looking in at him. He would have welcomed Strawser's help, but the door was locked. Knowing that he couldn't afford the time it would take to run over and let him in, Thomas decided to ignore Strawser and give his full attention to trying to save the infant gorilla's life.

As Strawser looked on in stunned amazement, Thomas performed mouth-to-mouth resuscitation on the baby gorilla for several minutes. Thinking he had done more than enough to revive her, he stopped for a few seconds to see if she would breathe on her own, but the baby continued to struggle. It didn't look like the gorilla was going to make it. Then, suddenly, after one, long, agonizing minute, she stopped hyperventilating and began breathing normally.

Thomas gently put the baby down among the towels and sponges he had used to clean her off and ran to the door. He told his friend that the baby gorilla was alive, but every second counted. Strawser must run to the zoo's administrative offices and find Davis and then track down a pair of hemostats. Thomas needed the clamps to pinch off the umbilical cord, which was still oozing blood. Despite all he had done for the gorilla already, he was reluctant to cut the umbilical cord himself. "I was only a second-year vet student and had never done it before," he says. With the baby breathing normally and the umbilical cord secured, Strawser and Thomas stood and stared at the little miracle. It had been a staggering few minutes, and the two young men were speechless. It didn't take long for Davis to arrive to see his new gorilla. As a wild grin spread across his face, he exclaimed, "I guess she *was* pregnant!"

Assuming Millie's gestation period would be the same as that of humans had turned out to be wrong. As is commonly known today, the gestation period of a gorilla is eight and a half months. After a 258-day pregnancy, Millie's baby had been born right on time.

Davis's wife, Mildred, was one of the first admirers to greet the newborn. Despite Mrs. Davis's excitement over the blessed event, years later she remembered being disappointed in the baby's appearance. She described the gorilla as all skin and bone, wrinkled and brown, with oversized fingers and toes attached to spindly arms and legs covered with black hair. But the gorilla's defining feature was her eyes. They were big, dark, expressive, and beautiful.

In the flurry of excitement, J. Wallace Huntington, the newly appointed chairman of the zoo commission, got locked into the great ape house. While everyone was paying attention to the newborn, Huntington had gone to check on the mother's well-being. In the process, he locked himself in the gorillas' quarters for more than an hour. It wasn't until a passerby heard Huntington calling for help that he was freed to join the excitement surrounding the newborn.

Before the initial shock had even worn off, the hoopla began. The gorilla's birth changed everything. Overnight, the Columbus Zoo went from being a typical small zoo that wasn't even considered one of the top two zoos in the state of Ohio to making headlines worldwide as a center of gorilla breeding.

One of the details Davis and his keepers shared was that Mac and Millie were being fed a steady diet of meat. Suddenly, zoos all over the world adopted this practice, hoping it would facilitate their own captive gorilla breeding efforts. The gorillas' meat-laden diet indicated just how little Columbus and the world's zoos knew about the animals: gorillas are vegetarians. Despite the concerted efforts of excited zoos determined to breed their own gorillas, it would be nearly five years before another gorilla was born at an American zoo: Tomoka, born at the National Zoo in Washington, D.C., in September of 1961. Five more years would pass before a third gorilla was born at an American zoo.

The fact that other zoos were soliciting advice from Columbus was a pleasant turn of events. Formerly skeptical zoo directors were now begging to see the newborn. "When we said she was pregnant, they all wanted to believe it," remembers Louis DiSabato, who was the zoo's curator of mammals in 1956. He eventually succeeded Davis as director of the Columbus Zoo and later spent twenty-six years as director of the San Antonio Zoo. "But how did we know [if it was true]? This had never happened before."

In Columbus, political officials were maneuvering for a

Robert W. Vesper and Louis DiSabato watch Colo sleep
in her first bed, a cardboard box padded with rags

position in the infant gorilla's spotlight. While he was spreading holiday cheer, M. E. Sensenbrenner, the mayor of Columbus, was also spreading gorilla cheer, handing out cigars to everyone. "It's a girl," the cigars read, as if the newborn were one of his own.

In a matter of hours after the gorilla's birth, the *Today* show was calling. Telegrams of congratulation began coming in. One of the telegrams read: "Congratulations to the Columbus Zoo on the birth of the baby gorilla. I hope it will live as long as I." It was from Bamboo, the thirty-year-old silverback of the Philadelphia Zoo. The *New York Times* ran daily updates on the baby gorilla's progress. *Time* and *Life* magazines made the baby famous in their pages. Thomas was invited to New York to tell his story on national television. Appropriately, he was a guest on the popular network game show *I've Got a Secret.* Host Art Baker wanted the gorilla on his television show *You Asked for It,* but her keepers didn't want to risk taking her away from the zoo.

There was no way for the zoo to prepare for its instant stardom, or for the immense responsibility that was involved in caring for its new celebrity. Without a real nursery, the infant spent her first few days, including Christmas morning, lying atop some rags in a cardboard box, next to the heater in the gorilla house boiler room. "We were all animal experts, but we were writing the textbook on rearing a baby gorilla. I can remember many times we looked at each other and said, 'What do we do now?'" DiSabato says. DiSabato himself had become a father earlier in the year, so his first impulse was to do what any concerned parent would do: he called his pediatrician. His son's doctor rushed over and gave the gorilla a thorough once-over, alongside the zoo's veterinarian. DiSabato's impulse was a good one. Pediatricians continue to be vital members of infant gorilla-care teams, even today.

The newborn was placed on a human infant formula called Olac (one part Olac and three parts boiling water) and given twenty-four-hour care by Mr. and Mrs. Davis, DiSabato and his wife, and other zoo employees and their families. On Christmas Eve, Davis excitedly told a newspaper reporter that the gorilla had "had a couple of jim-dandy burps" around midnight the previous night. It was a statement only a new father could appreciate.

The gorilla's much ballyhooed birth and all of the attention she was receiving around the Christmas holiday created a bit of controversy in Columbus. When a zoo employee erected a Christmas star over the zoo ape house to mark her arrival, some Columbus residents said it was sacrilege.

The greatest pressure on the zoo staff was simply to keep their priceless infant healthy. Their first medical scare came on Christmas Day. In her first three days of life, the gorilla's weight had fallen by nearly an entire pound, from an already frail three pounds five ounces to well under three pounds. The scare was short-lived, however, and by December 27 Davis was confident enough in the baby's health to send a telegram to his fellow zoo directors, inviting them to Columbus for a one-time-only viewing opportunity. He had received dozens of requests for an audience with the new arrival. "For her sake, we can't have people streaming in and out," he told the local newspaper. "We'll have them all at once."

Meanwhile, Millie Christina seemed to be recovering from the trauma of giving birth. She was throwing things toward Mac, returning some of the teasing and taunting the silverback had sent the expectant mother's way during her pregnancy. Millie's rediscovered spunk was a sure sign that she was on her way to a full recovery.

By the time the world was celebrating the arrival of 1957, Columbus's newest star was putting her toe in her mouth and scooting along the floor on her stomach. She was receiving all kinds of attention from curious and excited zoo workers, including a daily baby-oil rubdown. Her biggest medical problem was a nearly constant case of the hiccups. Davis ordered that she be fed more slowly, but her appetite was voracious, and trying to control the speed at which she ate was difficult.

While lying in the incubator that had been donated by Columbus Children's Hospital, the gorilla would show her

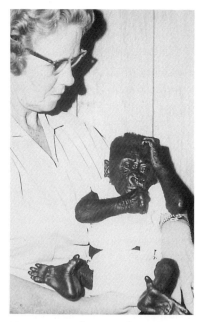

Nursery attendant Martha Gaines
holding Colo

strength by banging her teething ring frighteningly hard
against the incubator's glass. She also showed her dexterity by
loosening the lightbulb that hung over her head. From the moment she was born, her keepers kept detailed daily records,
monitoring her heart beat, respiration, and temperature.

The women who were providing care for the infant began
calling her typical baby names, like "Cuddles" and "Sweetie-
Face." Davis was quoted in the newspaper expressing disdain
for such pet names. "There's nothing official about that
name," Davis said. "I can't tell you what the name will be, but
it *won't* be Cuddles."

Since a public contest had been so successful before, when

the newborn's parents first arrived in Columbus, the city and the zoo decided to go to the people with the naming question. The *Columbus Citizen* newspaper ran a "Name the Gorilla Contest." The paper put up $25, to be awarded to the person who came up with the best name for the gorilla, and zoo commissioner Huntington sweetened the pot with $25 of his own money.

When Clark Gable heard the news of the gorilla's birth, and of the naming contest, he wired Walter Kessler, the manager of the Loew's Ohio Theater. The Hollywood star told Kessler he wanted to add a $100 savings bond to the contest, and the winnings grew to $150. Ironically, Gable had become enamored of gorillas during the filming of *Mogambo*, the movie Bill Said was working on when he died. Gable, who was born in Cadiz, Ohio, about a hundred miles east of Columbus, said that he was very excited that a zoo in his home state had been responsible for such a monumental event.

Nearly a month after the infant's birth, the name Colo was chosen from more than 7,500 entries reviewed by a panel of judges. Suggestions had arrived from all over the United States. The name Colo, a combination of the words Columbus and Ohio, had shown up on nineteen different entries, winning out over such suggestions as ColumBess, Gorilla My Dreams, Zoosie Q, Onyx (for onyx-pected), and Elvis. Mrs. Howard F. Brannon of Zanesville, Ohio, who had the earliest postmark on her winning entry, was awarded the prize money.

But Colo needed more than a name. She was growing in size and strength, and it was obvious that the zoo would need a special nursery for her. At forty-three days of age, on February 3, 1957, Colo weighed six pounds, was consuming about five hundred calories a day, and was gaining an ounce of

Colo demonstrating her trademark
spunk

weight every twenty-four hours. Davis knew Colo would
need better quarters, both to ensure her proper care and development and to handle the throngs of visitors who were
sure to line up at the zoo's gates to see her when the zoo
opened in the spring.

Davis went to Columbus City Council and asked for
$11,000 in emergency funds to build Colo a nursery. At least
one member of the council groused. "I wonder what they do
when these animals are born in jungles?" he asked. He suggested Colo be "rented" to the Detroit Zoo, which had offered to take her off Columbus's hands. Renting Colo was out
of the question, the other council members said. Finally, the
dissenting member gave in, saying he'd go along with all the
"ape hysteria."

Colo's new nursery was eighteen by twenty-five feet, with
floor-to-ceiling glass on two sides to allow for maximum

Visitors crowding around Colo's nursery

viewing by visitors. The room had perimeter heating so that the floors would always be warm; the nursery would be kept at a constant temperature of seventy-five degrees. Colo's space also had a sink, refrigerator, and stove for preparing her meals. The nursery was built as an addition to the new gorilla house, which had been completed between the time of Colo's conception and her birth.

Nobody could have known just how much of an impact Colo's presence would have on the tiny zoo. A well-known New York animal dealer said Colo's parents, Mac and Millie, would be worth $100,000 on the open market. Colo, on the other hand, was "bigger than money." Davis showed just how priceless she was by turning down an offer from Chicago's Brookfield Zoo to borrow the infant for $1,000 a day. "We could make a mint hauling her around," Davis told the media, "but she belongs right here."

More than a million people proved Davis right by visiting the zoo in 1957, shattering the previous attendance record. People of all ages pressed their noses against the nursery

window to catch a glimpse of the small gorilla. Colo's attendance record stood until 1992, the year the Columbus Zoo
displayed a pair of giant pandas on loan from China.

For the first six months of her life, Colo never left the sight
of her team of caretakers. Once Davis and his keepers were
confident that she was a healthy and growing infant, the zoo
hired a pair of nursemaids to take over her care. Thomas remembers Colo as a well-behaved infant who quickly became
spoiled by all the attention she received.

On her first Easter Sunday, Colo was dressed in ruffled
panties and a dress, and a picture was taken of her nibbling
on a daffodil. It wasn't unusual to see Colo wearing a dress
and hat, paraded in front of visitors and the local media like a
little princess. (Her keepers say her constant hat wearing as a
youngster has carried over into her adult life. She has the

Colo in one of her frilly outfits

Colo's first two teeth

habit of putting objects, such as food dishes, on her head, as if they were hats.) "I always hated that [we dressed her up]," Thomas says. "But all zoos did that with baby animals in those days. Thankfully, things have changed." DiSabato agrees that many decisions were made out of ignorance. "We know a lot more about gorillas today than we did then, and much of that knowledge can be attributed to the experiences we had with Colo. We documented everything," he says.

Today, keepers prefer that birth mothers raise their off-spring on their own. It is the natural thing to do, and it makes the most sense for the gorilla families. However, from the cir-cumstances Thomas observed when he found Millie and Colo, it was obvious that Millie wasn't willing or able to take the immediate and necessary steps to save and nurture her baby. Thomas said Millie appeared frightened by the birthing experience and attempted to abandon the infant. If the same circumstances occurred today, the infant probably would be removed from her mother and raised in a nursery until she could be reintroduced to her mother or placed with a gorilla surrogate and raised among members of a group.

Not only did Colo's birth bring notoriety for the zoo but it also placed the zoo's veterinarian in the spotlight. Like all zoological veterinarians of the day, Robert Vesper had no formal training in wild animal care. He had simply stepped forward and showed his willingness to apply his medical knowledge and knowledge of animals to the most exotic of the beasts housed at the zoo. Now, with Colo's birth, there was suddenly a great deal of added responsibility on Vesper's shoulders. Vesper's son Richard, now a veterinarian himself in the Columbus area, remembers his father taking the same baby scale his parents had used to weigh him and his siblings to the zoo for Colo. Richard also remembers taking many trips from the Vesper home in Upper Arlington, near downtown Columbus, about fifteen miles up Riverside Drive to the zoo just to check on Colo. It didn't matter where the family was going, or if a trip to the zoo was out of the way. For the sake of his father's peace of mind, a quick check to assure that Colo was all right became a family ritual.

Vesper had plenty of help with Colo's care. From the moment of her birth, Colo was one of the most studied animals in history. "Columbus is where it all started for gorillas," DiSabato says. "No matter what happens now or in the future, the zoo will always be remembered for Colo's birth." Zookeepers and scientists alike embraced the opportunity to monitor the development of the first gorilla born in captivity.

3

Watching Colo Grow

It seemed that everyone wanted to be part of the Colo success story. Scientists wanted to study her development; other zoos wanted to put her on display in their cities. Members of the zoo community and animal lovers everywhere simply wanted to catch a glimpse of the wonder child. Students of primate behavior were especially interested in the opportunity to determine whether gorillas could reach a higher level of intellectual development than their heavily studied cousin the chimpanzee. Scientists also hoped Colo could help them understand more about human learning, evolution, the nature of intelligence, and the role of environment in development.

Most comparative psychologists considered chimpanzees to be second to humans in intelligence, followed by gorillas. However, a study published in 1929 by the well-known primatologist R. M. Yerkes, entitled *The Great Apes, a Study of Anthropoid Life,* contended that gorillas were second to humans in terms of intellectual development. Yerkes based his finding, not on detailed research, but on the anatomical similarity of the brains of gorillas and humans. He argued that the gorilla brain would have a greater intellectual capacity than the smaller brain of the chimpanzee.

Because chimpanzees had been bred in captivity with great success, their development and learning capacity had been thoroughly studied and compared to that of humans. But comparing humans or chimps to gorillas was problematic, since no gorilla with a known birth date, known circumstances of birth, and known parents had ever been available for study. Comparative psychologists had studied mostly adult captive gorillas, with mixed results. Although gorillas generally scored lower than chimpanzees in intelligence, Yerkes claimed that lack of specific knowledge about the wild-captured gorillas contaminated the test data.

The husband and wife research team of Benjamin Pasamanick, the director of research at the Columbus Psychiatric Hospital, and Hilda Knobloch, the head of pediatric psychology at Columbus Children's Hospital, were the first to study Colo. Both scientists were on the faculty at Ohio State University's College of Medicine, and both had been on staff at the Yale University Child Guidance Clinic, directed by Dr. Arnold Gesell. Gesell had pioneered a developmental test for human infants and children that had been administered to thousands of children and was considered the standard measure of both motor and adaptive development. The Gesell test, and similar developmental tests, had been used on chimpanzees. Pasamanick and Knobloch wanted to use their experience to test Colo and determine how a gorilla's development compared to that of humans and chimpanzees.

The team applied for and received a research grant from the National Institute of Mental Health. With the zoo's permission, Pasamanick and Knobloch hoped to study Colo as long as her size and demeanor would allow—possibly several years. In their grant application they stated that, in the process of administering the Gesell test, they also hoped to test the theory that the longer a species spends in infancy, the more intelligent it will become as an adult. Their theory was based on the fact that a human, the world's most intelligent animal, spends the longest time in infancy. Colo would allow the researchers to see how quickly a gorilla matures, and how well the species develops specialized behaviors after infancy.

As part of their multifaceted study, the scientists also wanted to determine to what extent a gorilla could "unlearn" its gorilla ways. The team posed several questions, such as, "Could humans teach a gorilla to use its hands rather than rely on its natural instinct to use its feet?" Since chimpanzees had been successfully reared in homes and taught to understand multiple human commands, the couple hoped to determine if gorillas could do the same. Some of this research—such as attempting to teach a gorilla to act more like a human—would never be undertaken today. For the most part zoos do everything in their power to see that gorillas remain as true to their wild ancestry as possible.

Doctors Pasamanick and Knobloch requested that their study of Colo take place in a scientifically controlled environment, such as a laboratory at Ohio State, but that idea was quickly rebuffed. Colo was just too popular to be taken off display for any type of study, no matter how valuable the resulting data might be. The researchers also stated publicly that they hoped Millie and Mac could be bred at least two more times. They hoped one of the additional offspring could be raised with humans in a home, and the other left with its gorilla parents. Three gorillas living in three distinctly different environments would allow the scientists to measure the effects of environment on development.

The research team's aims, while scientifically valid, were unrealistic. Even if Mac and Millie successfully bred again, the zoo was not going to give up valuable animals for the sake of scientific study.

Despite the apparent exploitation of their prized gorilla, Davis and the zoo commissioners were willing participants in the study. They realized that it would bring added prestige to Colo and the zoo. Davis also thought the results of the study might help the zoo better understand gorillas, and therefore provide the animals with better care and enhance the zoo's chances at future breeding success.

The research team began studying Colo at the age of eighteen days. She was tested in her nursery on more than twenty occasions—at two-week intervals—before she reached her first birthday. She was also tested at fifteen, eighteen, and twenty-one months of age. In September of 1958 the team presented its findings at a meeting of the American Psychological Association in Washington, D.C., and later published the details of their study in the *Journal of Comparative and Physiological Psychology.*

In the first of their two reports, entitled "Gross Motor Behavior in an Infant Gorilla," Knobloch and Pasamanick reported that Colo was able to master basic activities such as sitting, toddling, walking, and standing about twice as quickly as humans and slightly faster than chimpanzees. From the time of the team's first visit, she showed accelerated motor development. She was able to bring her hands together in the midline of her body before she was three weeks old, whereas human infants usually cannot bring their hands together until the age of sixteen weeks. Colo was also able to lift her head while lying on her back from the very beginning—a behavior not attained by humans until twenty-eight weeks of age. At twelve weeks, Colo could hold her head steady and erect. At sixteen weeks, she was sitting, leaning forward on her hands. These developmental milestones don't occur in humans until twenty and twenty-eight weeks, respectively. Colo was standing while holding a railing at fourteen weeks. At eighteen weeks she was "cruising"—walking on two legs while holding a railing. She was secure in sitting at twenty weeks. All of these milestones were reached in less than half the time required by human babies. Knobloch and Pasamanick were able to make Colo walk on her two hind legs at thirty-eight weeks, even though, as nearly all gorillas do, she preferred to knuckle-walk on all fours. She would walk up stairs, with Knobloch holding her hand, at the age of forty weeks. Human toddlers usually don't walk until the age of one year and can't climb steps while holding an adult's hand until the age of eighteen months.

Although Colo's motor development was clearly accelerated compared to that of other primates, the scientists said this was not necessarily indicative of superior intellectual capacity. In fact, her accelerated development could prove their theory that a short period of infancy leads to a lower level of permanent intellectual development. It was her adaptive, or learning, behavior that would illustrate where gorillas stood on the hierarchy of primate intelligence.

The results of tests of Colo's adaptive behavior appeared in a second published report, which claimed her behavior resembled that of an autistic human. The research team pointed out that Colo's mother, Millie, had suffered from toxemia

Colo fingerpainting in the nursery

during her pregnancy and had had a convulsion two weeks before delivery. Since prenatal complications such as toxemia resulted in some neuropsychiatric disabilities in human infants, they asked, "In comparison with other gorillas, do we have evidence she is retarded and disorganized because her mother had toxemia? Has the factor of so much human association rather than helping her learn human traits, as one would expect, served as a further disturbing influence?"

Although their report stated that development of Colo's early adaptive skills was just as accelerated as that of her motor skills—about twice as fast as with human infants—she seemed to have reached a plateau at the age of six months. Her learning behavior was virtually the same for the next eighteen months; Knobloch never observed Colo engaged in such activities as stacking, scribbling, eating with a spoon, or pointing to objects she wanted.

The researchers' report also described the environment in which Colo lived, saying, for example, that Colo's supply of toys "would compare favorably with that of a well-equipped nursery school." The report detailed the layout of her nursery.

> Until she became too strong to manage, she was given the freedom of the very large room in which her glass-enclosed cage was located and had opportunities to explore sinks, stoves and refrigerators, as well as a rocking chair and pictures on the walls. She was held for her feedings, dressed in clothes that were either plain or fancy, as the occasion demanded, and when her diapers needed changing, was even removed discreetly from the admiring gaze of the multitude of visitors who usually gathered around the zoo's star attraction. Although she obviously developed a fondness for her caretakers, having a tantrum if they left her alone in her room or clinging fiercely to them if they attempted to put her into her cage against her will, she used human beings, by and large, as a means of attaining a desired object. The attitude of her attendants can be summed up by an overheard remark made by a disgusted teenager. "I think it's disgraceful, all this for a gorilla. Just think how much we would do if we provided all this care for our babies."

Knobloch and Pasamanick obviously agreed with the visitor. Looking back, Knobloch complains that the coddling of Colo sometimes got in the way of the research. On one occasion the keepers forbade Knobloch to perform a test that involved hiding a grape under one of three paper cups to see if Colo could find it: her keepers were afraid she'd find it and choke on it. On the other side, at least one keeper who was present at the time says Colo hated the testing. As if reacting to the presence of a veterinarian who regularly administers painful medicine, the sight of the researchers sent Colo into a screaming rage. Colo was said to have sometimes screamed from the moment they arrived until the moment they left.

Davis and the entire Columbus Zoo staff were infuriated by Pasamanick and Knobloch's second report. The suggestion that Colo might be disturbed or that her development might have been stunted by the care she was receiving in Columbus caused an uproar. "We who have been outwitted by Colo time and time again—I wonder where we'd rate in the I.Q. tests conducted by this team," Davis said. After the team reported its findings, an angry Davis denied the researchers any further access to the gorillas.

But although Davis vehemently disagreed with the scientific team's references to Colo's slow intellectual development, and the zoo's alleged contribution to it, he *was* worried that Colo was being spoiled by her caretakers. Like the researchers, Davis hoped that Millie would quickly give birth again—not to provide further subjects for scientific research, but so that Colo could have a gorilla playmate. Davis knew Colo was becoming too domesticated. She had never been allowed to see her parents, for fear that they would react violently to the sight of her. Colo only knew humans, and Davis knew it was important to create a balance between her existence as a gorilla and the influence of the people who were caring for her.

As everyone had hoped, Mac and Millie were mating again, and Millie began exhibiting all the signs of pregnancy a couple of months before Colo's first birthday. Millie was given a due date in the late spring of 1958. The keepers readied themselves for a second birth by gathering the best medical equipment available, and putting Millie on a round-the-clock watch. The same three nurses who were providing Colo's twenty-four-hour care took turns monitoring Millie. An incubator, oxygen, and an artificial respiration machine were set up next to Millie's cage. Since Colo had required mouth-to-mouth resuscitation immediately after birth, the keepers erroneously assumed that all captive-born gorillas have difficulty breathing during their first few minutes of life.

Two hundred fifty-eight days after Mac and Millie were observed breeding—the length of time it had taken for Millie to give birth to Colo—the watch around Millie intensified. Observers were waiting for Millie to lose her appetite or to exhibit the other unusual behaviors Thomas had witnessed the morning Colo was born. After ten more days had passed, they decided that the gorilla gestation period was indeed the same as humans, around 270 days, and, as many had suspected, Colo had been born prematurely. Even when day 275 arrived, Davis and his staff were still confident that Millie was pregnant and that Columbus would make history again with the world's second captive-born gorilla. However, by May 28, 285 days had passed, and there was still no baby. Millie had been under constant observation for nearly two months. Davis told a local reporter, she's "making a monkey out of me."

On June 1, Davis called off the twenty-four-hour watch. Millie had apparently had a false pregnancy. Veterinarians wanted to tranquilize her and give her a thorough medical exam to determine exactly what had happened, but it was too risky. Unsure of the proper dosage required to anesthetize her, they feared they might kill her. Although Millie would live a relatively long life at the Columbus Zoo, she would never give birth again.

When it became obvious that Colo was not going to get a sister or brother as a playmate anytime soon, Davis went to the zoo commission and told them he needed $5,000 to purchase an infant male gorilla from the wild. Colo's well-being depended on the zoo's finding her a playmate, he told them. Since Colo was worth ten or twenty times the amount of money they would have to invest for an animal from the wild, it would be a worthwhile investment. The commission agreed, but only if Davis could help them raise the money from private donors. Throughout the summer, Davis and the commission pleaded their case with Columbus businesses and individuals.

Meanwhile, an eighteen-month-old Colo was becoming playful to a frustrating degree. She learned to spring the latch on her cage, so it had to be padlocked. She often hung from the ceiling bars of her cage and delighted in dropping down on her attendants when they entered. Colo was also known to block the exit of her cage so her keepers couldn't get back out. One night, she spread apart the bars of her cage at the back of her nursery and walked down the hallway to the kitchen. Inside the kitchen area, a young female volunteer was feeding a baby bear. Colo wrapped her arms around the unsuspecting woman's neck and scared her half to death. It took more than an hour to get Colo back in the nursery. From that night on, Colo was caged, just like the rest of the gorillas.

By the end of the summer, the zoo had raised enough money to purchase a gorilla from the wild. Davis contacted Deets Pickett, a well-known American gorilla hunter based in

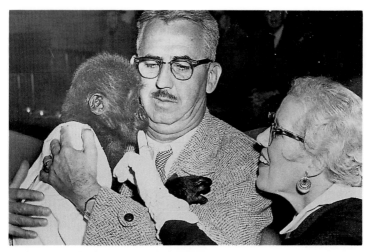

Earle and Mildred Davis with baby Bongo

Africa, to provide a male companion for Colo. Unlike Bill Said, Pickett did not gather his own animals. Instead, working from a base in the Cameroons, Pickett would "take orders" from zoos and then contact his network of local African hunters to make the capture. Pickett was a veterinarian and said to be among the most humane people in the gorilla-hunting business. Unlike his predecessors, Pickett worked hard to minimize the number of gorillas that were killed to make a capture. He was also successful at keeping the infant gorillas healthy after capture, feeding them canned milk and formula.

On October 1, amid a great deal of fanfare, a young male named Bongo arrived at Port Columbus in the arms of Earle Davis. Like Macombo before him, Bongo was named after the city nearest the site of his capture. Bongo had made the trip from Pickett's base in Yaoundé, Cameroon, via Paris, and an excited Davis had hand-carried the gorilla on an airplane

Colo, about two years old, and Bongo

from New York City. The only signs of Bongo's capture were a cut on his hand, which had been covered with a white cloth, and a frightened demeanor.

At the estimated age of eighteen months, Bongo weighed nineteen pounds. He was wide-eyed and gentle but seemed a little anxious: during a period of quarantine upon his arrival at the zoo, Bongo was spooked by a toy rubber alligator. The gentleness that would resurface in him later in life was apparent when he was young. His agreeable demeanor made it easy for keepers to change the dressing on his cut hand.

Bongo spent his first ten days at the Columbus Zoo isolated from the other animals. But it took Davis and his keepers only two days to put their new arrival on display for an excited public. Bongo would spend the next quarter-century on perpetual display. Davis told the press that the animal wouldn't so much as breathe on the zoo's priceless Colo until a complete set of tests could be run to determine if he had brought any parasites, bacteria, or viruses with him from the forests of Africa. Only after Bongo was given a clean bill of health did keepers schedule the young male's long-awaited meeting with Colo.

To prepare Colo for her first meeting with the gorilla who would be her mate for the next twenty-five years, she was dressed in a white dress decorated with red pinwheels and triangles. At the sight of Bongo, Colo leaped about her cage, climbing the bars and reaching between them to touch him. Bongo appeared to be more scared than excited, but he was interested enough to tentatively reach out to Colo. The zoo proclaimed the two to be perfectly matched, and the dream of a second generation of mating gorillas in Columbus was alive. By March 11, 1959, nursery records show the pair was sleeping peacefully together.

Bongo gave Colo the needed balance in her life between the human and the animal world. While she continued to be attached to her human keepers, Bongo became her principal playmate. The two were placed together sporadically and watched carefully every moment they were in the same cage.

As they matured, their health was monitored systematically as well. When the pair began teething, they gnawed on their wooden furnishings and walls. The splintered remains were ugly and dangerous, causing at least one gash in Bongo's head. In January of 1960, the zoo invested nearly $20,000 in tile-lined cages for the two young apes. Since Mac and Millie seemed unable to conceive again, Colo and Bongo were seen as the parents of the future gorilla generations in Columbus, and the zoo was willing to make a hefty investment to keep them healthy and happy.

Unexpectedly, in 1963, during a routine examination,

Colo, Bongo, Millie, and two orangutans—Jiggs and Maggie—were all diagnosed with a human strain of tuberculosis. Zoo veterinarian and soon-to-be Columbus Zoo director James Savoy was told by veterinarians at Ohio State University that the infected primates should be destroyed before they spread the disease to other animals. At any other zoo, this would probably have been done, but Colo's incredible value to the zoo and the city saved not only her own life but also the lives of the other apes. Instead of destroying the animals, Savoy chose to try to treat them, using a newly discovered human tuberculosis drug treatment. "Usually, when a whole collection is infected, the whole lot is gotten rid of," Savoy said at the time. "We are going about this in a different manner than before. A whole great ape collection has never been treated like this. They are, as far as possible, being treated and medicated like humans."

Warren Thomas had helped Savoy get his job as the Columbus Zoo veterinarian. When he left Columbus to open a new zoo in Oklahoma, Thomas had recommended his classmate for his vacated job. Savoy and Thomas stayed in close contact, relying on each other's expertise to treat their respective animal collections. When the Columbus ape collection was diagnosed with TB, Thomas was one of the first people Savoy called for advice. "He knew how valuable those gorillas were to the people of Columbus," Thomas remembers, "but it sure didn't hurt to remind him in no uncertain terms that it was his responsibility to do all he could to save them." A headline in the *Columbus Dispatch* read: "Colo Death Predicted in 90 Days." In the story, a "medical expert" from Ohio State stated that Colo, Millie Christina, and Bongo would all die from the disease within a few weeks.

Savoy relied on the expertise of doctors from the Tuberculosis Society of Columbus and Franklin County and veterinarians from the United States Air Force and Ohio State University to treat the animals. He determined that the primates had probably contracted the disease from one of their keepers, who had also come down with the disease. Because of the infection, Columbus became the first zoo to put plexiglass between the public and its gorilla collection. The glass was designed to protect the animals more than the public, since gorillas had been found to be susceptible to many of the same ailments and diseases as humans.

The TB scare came at a difficult time in the history of the Columbus Zoo. Earle Davis had died of cancer in 1960, ending his fourteen-year run as the zoo's superintendent. In April 1964 Stephen Kelly, who had succeeded Louis DiSabato as the zoo's director only one year before, was suspended in a controversy over $500 missing from the zoo's petty cash fund and asked to resign. Only eight years after Colo's birth, annual attendance had dropped to just 370,000.

A committee was formed to devise a master plan for improvements to the zoo that would restore attendance levels. During the committee's assessment of the facilities, conditions were found to be so bad that the option of selling off the animal collection and totally shutting the zoo down was discussed. "Most of the zoo buildings are insufficient to house the many animals now in the collection," the report said. "Compared to other zoos throughout the country, the Columbus Zoo can be best termed as 'poor.' "

The zoo veterinarian James Savoy was named to succeed Kelly as director. Savoy worked hard to turn things around. During his time as the zoo's veterinarian, he had complained bitterly about the deplorable conditions in which the animals were forced to live. One of his early successes as zoo director was being able to report that the TB that had appeared in the gorillas had been arrested by medication. Savoy said a series of skin tests, chest X-rays, and a sampling of stomach fluid showed only a mild positive existence of TB in Colo, Bongo, and Millie, although the tests showed a definite positive in the two orangutans. Savoy announced that the gorillas would continue to receive medication but that the infection had been stopped. It was wonderful news for the zoo and allowed the continuing expansion of Columbus's famous gorilla collection.

Around this same time, Daniel Galbreath, one of the city's most well-known businessmen, became interested in the zoo. His advocacy of the zoo and his support of its growth and prosperity would continue for decades. He was a big-game hunter himself who kept a menagerie of wild animals on his horse farm in southwest Columbus. In addition to champion thoroughbred horses, he owned major league baseball's Pittsburgh Pirates. M. E. Sensenbrenner, another future staunch supporter of the zoo, especially of its gorillas, also entered the picture in 1964, when he was reelected mayor of Columbus. The beleaguered zoo got a further boost in 1967, when the first zoo bond issue was passed by county voters. Most of the $2.5 million was spent on infrastructure, such as sewage and drain pipes.

A year later, Colo and the gorillas would again put Columbus's zoo in the spotlight. Colo was apparently pregnant. Just after her eleventh birthday—the same age as her mother had been when she gave birth—Colo began gaining weight. Her keepers had observed Bongo and Colo breeding the previous spring and summer and had hoped for the best. After their experience with Millie a decade earlier, however, they had made no public announcement and simply kept watch. According to their calculations, Colo was due in March. She was pulled from her cage with Bongo on January 3, just in case she gave birth early. Keepers wanted her to give birth in isolation, fearing Bongo might injure or kill the baby. Paper was hung on the outside of the viewing glass at the front of Colo's cage to afford her some privacy from the public's view.

Although the keepers tried to make her as comfortable as possible, lights shined in Colo's cage twenty-four hours a day, in case the birth occurred at night. Keepers thought switching on the lights to check on her at night might spook the expectant mother, so they simply left the lights on. (Today, keepers and veterinary staff often use flashlights to monitor the progression of a birth.) On Wednesday, February 1, 1968, about 3:00 in the afternoon, Colo and Bongo became parents. Although the newborn arrived more than a month earlier than expected, the zoo was ready. Keepers say Colo was in labor just fifteen minutes. Her baby was the first second-generation gorilla to be born in captivity and only the eighteenth captive-born gorilla in the world. "As soon as it was born it cried. It sounded just like a baby's cry," Savoy told reporters. "And, with the sure sign of life, I began to click my heels with joy."

Colo acted like a loving mother in the first hours and days of her baby's life, even though she had never been cradled or fed by her own mother. As keepers looked on soon after the baby's birth, Colo picked up her child and held it in front of her nose and simply gazed at it. The zoo had readied an incubator for the infant, and Savoy and his staff had expected to pull the baby and raise it in the nursery. But Colo's tenderness with the child convinced keepers to leave the baby with her mother as long as it remained healthy. It was an unusual risk to take during an era when all captive-born gorillas, owing to their tremendous value to zoos, were immediately pulled from their mothers and raised in nurseries.

As the baby's first couple of days passed, Colo continued to gently cuddle and cradle the infant. Keepers hoped the baby, whose sex was still unknown, would begin nursing. Although the newborn appeared to be attempting to nurse, veterinarians were concerned that Colo had not produced a sufficient amount of milk to nourish the child. Keepers fed Colo milk in a tin cup, hoping it would stimulate the production of her breast milk. However, by Sunday, February 5, keepers had not observed a single episode of successful nursing. Fearing that dehydration and malnourishment might endanger the baby, Savoy and his keepers decided to pull the baby from Colo.

The baby was a girl, weighing about four pounds. As before, a contest was held in Columbus to choose her name. On March 14, 1968, she was given the name Emmy, in honor of Mayor M. E. Sensenbrenner, who said he "felt like a grandfather." The name was submitted by a ten-year-old girl. An excited public got to see Emmy for the first time on April 6, 1968.

But a year later, a cloud of bad news again hung over the zoo's ape house. Routine examinations of the gorilla family found a recurrence of tuberculosis. Just as in 1963, the prognosis was grave. Luckily, Savoy and the animals again proved the doctors wrong. None of the animals succumbed to the disease, although the chilling prognosis created a renewed sense of the animals' value to the zoo and to the community. Over the years, the threat of tuberculosis refused to go away and would continue to influence decision making relating to the zoo's gorilla collection.

Just months after doctors said her life was in peril from TB, Colo gave birth to her second child. At 1:40 in the morning on July 18, 1969, Oscar arrived, weighing nearly five pounds at birth. This time, fearing her infant would be exposed to TB, the keepers sedated Colo and pulled her baby from her cage just minutes after she gave birth. From her cage in the nursery, eighteen-month-old Emmy could see her new brother lying in his incubator. "We may have a little problem with Emmy. Like many a human sister, she's a little jealous of the new baby," Savoy told local reporters. Meanwhile, Mayor Sensenbrenner proclaimed that he was a grandfather again and began campaigning for the newborn male to be named "Sensy." "That's a boy's name, isn't it?" the mayor asked rhetorically. The mayor didn't get his way. Another naming contest resulted in the name of Oscar. With a little help from her mother, a seven-year-old girl wrote: "Because Colo has nothing but winners, her little boy should be named Oscar, to go along with her Emmy."

Oscar's birth was such an event in Columbus that the news of his arrival was given as much space on the front page of the *Columbus Dispatch* as the flight of Apollo 11, which had lifted off on NASA's first manned mission to the moon. When Oscar was two days old, fellow Ohioan Neil Armstrong walked on the moon.

Shortly after her fifteenth birthday, Colo's life of surprises continued. Three days after Christmas 1971, she unexpectedly gave birth to her third child, a three-pound fourteen-ounce female. The birth went remarkably well considering that nobody at the zoo even suspected the gorilla was pregnant. "I hate to admit it," Savoy told the press, "but we had no idea she was pregnant. There was no development to indicate she was pregnant." Colo had exhibited a false pregnancy the year before, and Savoy and the gorilla keepers thought the few indications of pregnancy she had shown were merely part of her false pregnancy pattern.

Like her brother, Oscar, Toni, so named to complete the trio of "award" names for Colo and Bongo's children, was pulled immediately from her mother to protect her from exposure to tuberculosis. The Columbus Zoo nursery was now full of Colo and Bongo's children. Toni joined Emmy, who was nearly four years old, and Oscar, aged seventeen months. The zoo's gorilla breeding program was now so successful that it was creating a problem. Savoy and his keepers didn't know where they were going to put so many gorillas. Since the adults were all thought to have been exposed to TB, Savoy thought it was too risky to put the three young gorillas in the same building with them. Either another gorilla house had to be built, or the animals would have to be moved off-site. The Detroit Zoo was eager to have Emmy, Oscar, and Toni and made a generous offer to purchase the three young gorillas. But Savoy turned Detroit down and instead converted some vacant space in the zoo's former hospital to house the three animals. It was a temporary solution at best, but it kept the gorillas in Columbus.

In February of 1976 the zoo released elaborate plans for a $4 million gorilla habitat to be built with private donations and about $1 million set aside from a recently passed bond issue for the zoo. The plans, drawn up by a Columbus architectural firm, promised to provide space for twenty gorillas. The new habitat would be located prominently near the zoo's entrance. However, the necessary funds for the project could not be raised, and the plans were scrapped.

Colo experienced another close call in 1974, when Savoy became concerned that she hadn't had a bowel movement in some time. Gorillas eat a naturally high-fiber diet, so constipation is not usually a problem unless there are extenuating circumstances. Savoy and the keepers did their best to keep Colo as hydrated as possible, but nothing seemed to be working. Fearing the constipation might be a symptom of something worse, Savoy called Harrison Gardner, a well-known veterinarian at Ohio State who had had a great deal of experience working with large farm animals. The willing and congenial Harrison agreed to come to the zoo to look at Colo. It would be the beginning of a fourteen-year relationship between the zoo and Gardner.

By the time Gardner arrived, Savoy had already decided they were going to have to anesthetize Colo and take her to the university for a full examination and an enema. Although he had anesthetized large farm animals before (known by vets and zoo keepers as "knocking down" an animal), Gardner had had no experience with anesthetizing a gorilla. Savoy helped Gardner determine the proper dosage necessary to tranquilize Colo, but after taking two darts full of the drug, Colo seemed unaffected. Knowing that the darts sometimes didn't deploy the medicine or that they hit muscle and didn't get the medicine into the bloodstream, Gardner prepared a third dart. "Don't you kill my most valuable animal!" Savoy said to Gardner as he took aim. The third dart put Colo into a deep sleep. When Gardner entered the cage and reached Colo, however, she had stopped breathing. Luckily, Gardner was able to bring her around with artificial respiration.

During the next seven days, Gardner had to knock down Colo three more times to administer enemas before she was finally able to move her bowels. After a complete examination, the veterinarian determined Colo's severe constipation had been caused in part by eating too many unpeeled bananas. Even though Gardner had many more anxious times as the zoo's veterinarian, he says that first experience with Colo was one of the most trying and difficult of his career at the zoo.

Although unfortunate circumstances had led her keepers to pull Colo's three babies shortly after their births, her maternal instincts remained strong. In 1987, at the age of thirty-one—an age when many captive gorillas are doing little more than living out their years—Colo was given the opportunity to be a provider for a gorilla infant. Through the Columbus Zoo's innovative surrogate program (discussed in detail in chapter 9), Colo took over the mothering duties for her fourteen-month-old grandson, J.J., after his mother, Toni, was unable to care for him. Colo turned out to be an excellent mother. Not only did the arrangement enrich J.J.'s life but it also seemed to help Colo. During her period as a surrogate, keepers say Colo showed overt ovulatory behavior for the first time in years. Being a provider also elevated her social status in the gorilla troop, and she was calmer and easier to care for. After having three offspring and fifteen grandchildren, Colo was finally given the opportunity to mother an infant herself. The zoological community had been benefiting from observing Colo all her life, and the successful introduction of J.J. and Colo provided yet another important lesson. The keepers came away from the experience convinced that their surrogate program would work.

Colo carrying her grandson—and surrogate son—J.J.

In 1996 space again became an issue when construction was scheduled to begin on a new gorilla habitat. The keepers decided to temporarily send one of their three gorilla troops to the Detroit Zoo, which had not had gorillas in more than fifteen years. Silverback Sunshine's group, which included Colo, was chosen to move to Detroit, but the Columbus keepers didn't want to send Colo away at the age of thirty-nine, even though she was quite healthy. The keepers decided to introduce her to a new group and send another female, named Cora, to Detroit with Toni and Sunshine. Not only would this allow Colo to stay in Columbus but it would also get her away from Sunshine, who her keepers say made her uncomfortable.

It was decided that Colo would be introduced to Annaka, a young male who had arrived in Columbus in 1993 on loan from the Philadelphia Zoo. At the age of ten, Annaka had not reached silverback status. He was a rambunctious blackback whom keepers considered a "silverback in training," likely the next in line as a Columbus troop leader. The keepers thought Colo's social savvy and status might help Annaka mature. And, after all she had been through over the years, they were confident she could hold her own against the young male.

Today, even at the age of forty, Colo has not lost any of the spunk she showed as an infant. In addition to being accurate when throwing objects, she is also known as a dead-on spitter. "Granny," as some of the keepers call her, often rains saliva down on people she is trying to test, like new keepers.

The zoo's veterinary staff says Colo is free of the tuberculosis that once was thought to be life threatening. In fact, some of the vets, including Gardner, have privately wondered if the positive tuberculosis tests were accurate. Gardner has performed the necropsies on Columbus gorillas once suspected of having tuberculosis and has never found any traces of the disease. In fact, TB—except for a cold-blooded strain found in a snake—has never been found in any Columbus Zoo animal. It would be ironic if the apparent presence of tuberculosis, which dominated the zoo's treatment of its gorillas for many years, was nothing more than a series of false test readings.

Today, Colo's only signs of age are a bit of graying around her distinctive face and a mild case of arthritis. She is kept on a special diet, which includes a protein drink similar to the drinks prescribed for human senior citizens. Surprisingly, her keepers say she has not yet reached menopause.

The Columbus keepers are in awe of Colo's intelligence. They cite example after example of her ability to observe and learn. Keeper Charlene Jendry often shares two Colo stories that she says exemplify the gorilla's intelligence. Both involve "contraband" that Colo has found in the zoo's outdoor habi-

J.J. grooming Colo

tat. Although the keepers routinely examine the habitat before the gorillas are let outside, small objects are difficult to
detect in the grass, which is intentionally kept long to encourage the animals to forage. One summer day, Jendry says she
was informed that Colo had found a set of children's plastic
keys in the grass. The keys posed enough of a choking hazard
that the keepers wanted to get them away from Colo. Thinking that an enticement of food would be necessary to "make a
trade" with the gorilla, Jendry went to the Wendy's fast-food
restaurant next door to the ape house on the Columbus Zoo
grounds and picked up a chocolate "frosty" ice cream treat.
The ice cream got Colo's attention, and she walked to the
mesh carrying the keys. As Jendry talked to Colo, pointing
to the keys and explaining "the deal," Colo placed the keys

under one of her feet, out of Jendry's reach. As Jendry offered a spoonful of ice cream, Colo broke the plastic key chain and offered Jendry one key in return. It took a half dozen bites of ice cream to get all the pieces of the keys from Colo.

The other incident involved a much more dangerous piece of contraband: a length of metal chain, apparently left behind by some workers who had just completed routine repairs inside the habitat. Jendry feared Colo might swing or throw the chain, potentially injuring herself, another gorilla, or a zoo visitor. Once again using food as her negotiating tool—this time it was yogurt—Jendry says Colo offered a few links of the chain through the mesh in return for the first bite of food. Jendry knew if she went for the chain, Colo would forcefully pull it back. With each additional bite of food, Colo pushed more of the chain through the mesh, not relinquishing the entire length until the yogurt was gone.

In November of 1984, Colo almost freed the entire gorilla collection. After a keeper apparently left Colo's cage door ajar, she escaped into the keeper area. Her first stop was Bongo's cage, where she turned the lever to the airlock, just as she had watched her keepers do for so many years. But the lock didn't immediately release. Had Colo showed a little patience and held the lever down longer, Bongo would have been free. After unsuccessfully trying the lever several times, Colo decided to move on. She checked sinks and garbage cans, and even picked up the telephone and put it to her ear. The keeper put out the word: they had a "code 1A in the gorilla house," an escaped gorilla. Although she was confined to the keeper area of the gorilla house and had not entered any public areas, zoo visitors were asked to wait inside the zoo's reptile building as a precaution, in case Colo figured out a way to escape the confines of the ape house. Other keepers, armed with tranquilizer guns, guarded the exterior door. About an hour after her escape, zoo veterinarian Gardner climbed in the rafters of the ape house and got a clear shot at Colo. He used a blowgun to hit her with a tranquilizer dart. In moments, a drowsy Colo voluntarily returned to her cage, where she went to sleep.

A week after her fortieth birthday in December 1996, Colo escaped again, walking through a transfer door left open as keepers cleaned an adjoining holding area. Zoo executive director Jerry Borin says Colo saw something different in her routine and took advantage of it. Although she was loose in the keeper aisle, again she didn't reach the outside. She was lured back to her cage with some sweets; no tranquilizer gun was necessary.

By all accounts, Granny can be a grump. Warren Thomas says he hasn't seen Colo in more than twenty years, but the last time he visited her she tried to bite him. However, longtime keeper Beth Armstrong says she can also be an incredibly pleasant animal to be around. Armstrong appreciates her nerve, intelligence, and beauty. Her keepers celebrated Colo's fortieth birthday by offering the world's first captive-born gorilla her favorite foods: kiwi fruit, grapes, and multigrain cakes.

Today Colo's hair is still mostly deep brown, and she still has the large eyes and long eyelashes that charmed her earliest human companions into calling her Cuddles.

4

Bongo, and the Expanding Columbus Gorilla Program

Ongoing study of Colo taught the Columbus Zoo and the entire zoological community a great deal about captive gorilla breeding and development. Warren Thomas says Colo's birth was to zoos what the breaking of the four-minute mile was to world-class runners. After decades of striving for a seemingly unattainable goal, somebody had finally achieved it. And once it was proved that the goal could be reached, many others gained the confidence to follow. Nonetheless, Colo's birth did little to change the zoological community's approach to the care of captive gorillas. Decades would pass before zoos got beyond the standard practice of simply keeping gorillas fed and their cages clean—whether isolated or in pairs—and putting them on constant public display.

Breakthroughs in captive gorilla care would not come until gorilla behavior was successfully observed and studied in the wild. Field biologist George Schaller was the first American to go to Africa for the purpose of modern scientific observation of gorillas. In 1958, before making his first trek to Rwanda, Schaller spent several days with Thomas at the Columbus Zoo. Although he was not yet thirty, Thomas, because of his experiences with the world's only captive-born gorilla, was considered one of the country's foremost gorilla authorities. At the end of his visit to Columbus, Schaller asked Thomas if a field biologist would ever be able to get close enough to gorillas in the wild to actually study them. "Not a chance," Thomas remembers telling him. "But I wish you the best of luck anyhow."

Schaller and his wife, Kay, in fact became the first scientists to make peaceful contact with gorillas. Schaller's scientific work—later published in *The Mountain Gorilla: Ecology and Behavior; The Year of the Gorilla;* and, with Michael Nichols, *Gorilla: Struggle for Survival in the Virungas*—went a long way toward changing the public's perceptions of gorillas and destroying the long-standing myth that they were ferocious creatures who had to be approached with a weapon at the ready. "All I wanted from them was peace and proximity," he wrote. He was given both, and the world knows a great deal more about the animals as a result. In 1967 Dian Fossey followed in Schaller's footsteps and expanded upon his studies. Yet even after the biologists' findings about gorilla behavior—specifically their social behavior—were published, zoos were slow to apply the paradigm of the wild to their captive gorilla programs.

When future head gorilla keeper Dianna Frisch joined the Columbus Zoo staff in 1976, Colo and Bongo had been together, and on constant display, for nearly twenty years.

The original great apes building, where the animals were on constant display

Their days consisted of little more than eating, sleeping, and occasionally breeding. According to Frisch, the pervasive attitude of the era was "If you give them nothing else to do, gorillas will breed." Although Bongo had sired three offspring with Colo—perhaps reinforcing the attitude of the day—it was obvious to Frisch that the unstimulating environment was unhealthy. Every movement the gorillas made within their stark concrete and steel cages occurred in plain view of their keepers or the visiting public. There was no place for them to hide.

This constant attention seemed to particularly irritate and anger Bongo. He was often rough with Colo—too rough in the eyes of the keepers—and would sometimes force her to give up her share of food at feeding time. When Colo did manage to get something to eat, keepers often observed Bongo making Colo regurgitate her food for him to eat. Bongo also directed his apparent anger and frustration toward his keepers. He threw things at them, including feces. Frisch says there seemed to be a constant feeling of tension—for both the animals and the keepers—in the Columbus Zoo gorilla house.

Keeper Beth Armstrong feels that Bongo in fact "represents everything we did wrong in zoos." She points out that he lived most of his life in a concrete cage without bedding. Though he had access to a small outdoor cage, he was still on display when he was outside. His offspring with Colo were all pulled and nursery reared, robbing him of the experience of having young ones among the group, which Armstrong says is the ultimate stimulation for virtually all species. Bongo's aggravation at the public became his only real outlet. "I would look at Bongo and say, 'This is absolutely wrong,'" Armstrong says. "I channeled my frustration about Bongo into a positive thing. He made me determined to change things."

It wasn't until a 1983 visit to the Columbus Zoo by Dian Fossey that things truly began to change. Fossey was on a promotional tour to support her best-selling book, *Gorillas in the Mist.* The keepers were all anxious to meet Fossey but privately feared that she might react negatively to the tense atmosphere in the gorilla house. When Fossey entered the building, Bongo was lying down in his cage, his back to the ape house entrance. Fossey saw him and leaned down at the opposite end of the cage and began vocalizing to him. Frisch says Bongo leaped up and immediately moved across the enclosure to Fossey. Frisch, Armstrong, and keeper Charlene Jendry watched in amazement as Bongo and Fossey spent the next several minutes communicating, literally nose to nose.

From left, Charlene Jendry, Dian Fossey, Beth Armstrong,
and Dianna Frisch during Fossey's 1983 visit

Although she had made her negative feelings about zoos
and animal captivity known in her book, one of the world's
foremost primatologists made a strong connection with the
three gorilla keepers and the Columbus gorillas. She ended up
extending her stay in Columbus and offered several suggestions that dramatically changed the way gorillas were cared
for at the zoo. Fossey said the gorilla program should focus on
providing for the animals' social and psychological needs as
well as their physical care. For instance, she suggested that the
zoo's practice of keeping the gorillas healthy by making sure
their cement-floored cages were completely clean and devoid
of debris was unstimulating. Fossey suggested that instead the
gorillas be provided with hay so that they could build nests, as
they do in the wild. The hay would provide both comfort and
stimulation. Heeding her suggestion, the keepers watched in
amazement as the animals who had been brought to the zoo
from the wild—namely, Mac and Bongo (Millie had died in

1976)—built elaborate nests. In fact, Mac spent hours working on his nest, concentrating on it until every detail was perfect. Colo, on the other hand, wasn't sure what to do with the hay. She spread it in a ring around her body but didn't put any of it beneath her. It was an amazing display of a learned behavior that only the animals from the wild seemed to understand.

Fossey also suggested that the keepers cut the gorillas' food into small pieces and hide it in the hay, because gorillas like to forage for their food. Not only did the keepers begin hiding food in the hay inside the animals' cages, but they also let the grass grow in the outdoor habitat, so food could be hidden there. In addition, Fossey introduced the keepers to some basic gorilla vocalizations. Understanding the gorillas' grunts and coughs allowed the keepers to communicate with the animals in basic ways and to comprehend when they vocalized with each other. "These were changes we could make immediately," Jendry says. "[Zoo director] Jack Hanna had created an atmosphere that empowered the keeper staff, and allowed these things to happen without being 'committeed' to death."

On the surface, Fossey's suggestions may seem simple and straightforward, but in truth they were revolutionary. They ushered in a new way of thinking at the zoo—one that had keepers focusing on the way gorillas live in the wild. And for the first time, the keepers began to recognize the importance of a proper social structure within gorilla troops. This discovery led them to create a program focused on the mother-rearing of infants and on age diversity within gorilla groups. The keepers came to realize that their program could no longer be judged simply by the number of gorillas born at the zoo;

rather, their program's success or failure would be determined by the quality of the gorillas' lives after birth. It became clear that it was in the best interest of the next generation of captive gorillas for the staff to make a total commitment to the early socialization of all gorilla infants, especially those that had to be raised in the zoo nursery. The keepers hypothesized that hand-reared gorillas would have a better chance of raising their own young if they were introduced to other gorillas at an early age. Having seen hand-reared females who couldn't or wouldn't properly raise their own offspring, the keepers wanted to end as far as possible the cycle of hand-raised animals. They recognized that the blame for maternal neglect might not lie with the animals but with the keepers, who hadn't properly prepared them to handle the responsibilities of parenthood.

The keepers knew even before Fossey's visit that the zoo's care of the gorillas was inadequate; Fossey simply confirmed the need for major philosophical changes. Within a year of her visit, Columbus's revised approach took a giant leap forward with the opening of a new $700,000 gorilla habitat. At a time when the trend in zoos was to build naturalistic exhibits that attempted to replicate the look and feel of the wild, Columbus decided to go in an entirely different direction. In keeping with its social group philosophy, the zoo chose an exhibit design based on John Aspinall's "gorilla villa" at the Howlett's Zoo in England. Aspinall had successfully housed large groups of gorillas using the concept of the "villa"—a round, several-story structure surrounded by mesh and featuring climbing areas, ropes, and varied terrain. Aspinall had been an important proponent of gorilla socialization, stating publicly that isolated gorillas were "virtually dead," and that even

The "gorilla villa"

paired gorillas were basically "non-animals." He considered isolation a form of punishment, akin to solitary confinement for humans. "You have a problem," he told a gathering of American zoo directors in Columbus in the fall of 1985. "Your zoos are built for people, and not for animals."

Columbus's new "gorilla villa" addressed Aspinall's complaint about zoo exhibits. Not only did the renovated habitat give the animals a stimulating outdoor play yard, the new facility was larger inside, too, and featured simple but important stimulants including hay, ropes hung from ceiling beams, and an old tire. Perhaps more important, the new arrangement allowed the gorillas to be "off display."

Although a small group of gorillas had been given the opportunity for privacy earlier, in 1979, when the zoo had converted a former elephant yard to an outdoor gorilla habitat called Oscar's Yard, Bongo, Colo, and Mac had remained in the old ape house. Now that the new habitat was built and

Bongo was given the option, he often chose to stay inside, away from the public's view. According to the keepers, the renovated facility decreased the animals' tension. Bongo, especially, seemed to be more comfortable. He became even more playful, and his "good days" became the rule, rather than the exception. The opportunity for real activity, and a more sensible diet, also allowed Bongo to lose a great deal of weight. At 425 pounds, he was a nearly perfect physical specimen. Armstrong says the slimmed-down Bongo was an exceptionally handsome gorilla. "He had sheer physical beauty," she says. "His heavily lined face and perfectly proportioned body were extraordinary."

Bongo's former weight problem was not unusual. In fact, the problem of sedentary and overweight gorillas was pervasive in zoos in the 1960s and 1970s. However, as more was learned about gorillas' dietary needs and as larger spaces were provided for them, the situation improved. Bongo's diet, which in the early days featured meats and sweets, was changed to include a breakfast of fruits (grapefruits, apples, and bananas) and vegetables (lettuce, spinach, cooked sweet potatoes, and carrots), topped off with a hard boiled egg and vitamins. For lunch, like today's gorillas, he received yogurt and a special protein drink developed by the keepers. Throughout the course of the day, he had sunflower seeds to snack on, which provided not only nourishment but also stimulation and dexterity exercise as he shelled the seeds with his hands and teeth. Like other gorillas, Bongo also snacked on a prepared animal food product known as monkey biscuits, and he peeled and gnawed on willow branches throughout the day. At times, he and other animals were treated to some sugarcane, coconut, or watermelon. (One summer, the word got out that the gorillas loved watermelons, and the zoo

Bongo in the gorilla villa

was flooded with so many donations from farmers and gorilla lovers that the gorillas couldn't possibly eat them all; zoo employees were taking them home for their families.) For dinner, the gorillas were often fed bok choy and endive, more sweet potatoes and carrots, onions, fruit, and a white potato. Bongo also had a taste for roses. One particular docent grew pesticide-free roses in her garden especially for Bongo. With the permission of the vet staff, keepers would de-thorn the plants and hand them through the mesh to Bongo, who delighted in munching on them.

Bongo's intelligence, presence, and newly calm demeanor made him a favorite among the keepers. Frisch says that, unlike any other gorilla she has ever worked with, Bongo had a way of making a deep impression. "He could get into your mind," she says. Bongo also made an impression on Fossey, who regularly asked about him in letters she exchanged with the zoo. Armstrong says she used to go home at night and lie awake trying to figure out ways to improve Bongo's life. The keepers would sit for hours simply watching him. Frisch and others kept journals of his activities, but, they say, they watched him for pleasure as much as for scientific observation. Frisch and Bongo often played a game involving a peanut, hidden in one of her hands. She would have Bongo pick which hand the peanut was in. She would also add a marshmallow to the game and let Bongo select his favorite. He always ended up with both.

One night, Frisch remembers Bongo was in the building's north cage while she was sweeping up in the area. The light was out in the hallway. Frisch complained out loud. "The light is out, and I can't see a damn thing," she said. Perhaps reacting to her frustration, Bongo smacked the wall, which triggered the burned-out bulb to flicker back on again. An amazed Frisch says it was like Fonzie from the *Happy Days* television show smacking the jukebox and making a song begin. Like the Fonz, Bongo was the cool guy whom everyone liked. And all hoped that Bongo shared their affection.

Bongo once represented the zoo's gorillas on a consumer advocate television program, testing a briefcase that was supposed to be able to withstand the strength of a gorilla. Hanna agreed to let keepers place the luggage in Bongo's cage, saying it would be a harmless form of stimulation for Bongo. Moreover, Hanna always believed in the power of publicity; certainly this demonstration would attract media attention. Bongo's bout with the briefcase did become a big story, although some other zoo directors questioned Hanna's judgment for allowing one of his prized gorillas to be exploited in a stunt that might get him hurt. In fact, Hanna had told the TV crew in no uncertain terms that if there were any sign that Bongo might get hurt, he would bring the demonstration to a swift end.

When the briefcase was presented to him, Bongo examined and sniffed it in typical curious fashion, as any gorilla would. After playfully pushing it around for a few minutes, he amazed observers by popping it open at the latches, using his thumbs, just the way it was designed to be opened. The inside offered new textures and smells, which occupied Bongo for a few more minutes as he quietly examined it. Then he swiftly lifted the briefcase with both hands, brought it up in front of him, and bent it backwards against its hinges. He bent it back and forth against its hinges a few more times until it was completely mangled. He finished off the toughness test by hurling it against a wall, which broke the briefcase in two.

As a more natural form of stimulation, Frisch decided that Bongo might enjoy meeting a live rabbit. As she predicted, Bongo was enthralled by the small white bunny. When Frisch brought the rabbit over to his cage, Bongo got down on his elbows and studied it. Then, he reached his huge finger through the mesh of his cage and gently stroked the animal. After touching it, he offered his own hand to the rabbit, inviting the animal to touch him back. Frisch says it was a great example of the gentleness belied by Bongo's intimidating size.

Frisch decided that the next step in freeing Bongo of the longtime inadequacies of his care would be to end his twenty-five-year one-on-one relationship with Colo. After so many years together, the two had become uninterested in playing and mating. By 1983, thanks to the field research conducted by Fossey and others, the Columbus keepers knew that male gorillas had multiple partners in the wild. Not only was it the natural thing to do, but Frisch thought that separating Bongo and Colo would provide each with a more stimulating environment, upgrade the quality of their lives, and increase their opportunities to mate.

When the two were separated, Colo and Bongo reacted differently: Bongo called for Colo almost constantly, while Colo seemed indifferent. To keep him from calling to her day and night, the keepers had to hang cloth between the mesh of the cages so that Bongo couldn't see her. He thwarted their efforts by climbing on a tire in his cage to see over the cloth and catch a glimpse of Colo. In an effort to get Bongo's attention away from Colo, the keepers introduced him to Donna, a seventeen-year-old female on loan from the Como Zoo in St. Paul, Minnesota. Next they tried Lulu, a twenty-year-old female on loan from the Bronx Zoo. Finally, they introduced him to Bridgette, who keepers had decided was a perfect match. Bridgette was a slightly overweight and even-tempered animal. At the age of twenty-five, she was a proven mother who had had seven offspring, including three, sired by Oscar, born in Columbus since her arrival on a breeding loan from the Henry Doorly Zoo in Omaha, Nebraska. After the two were put together, Bridgette aggressively pursued Bongo, and he was receptive to her advances. It seemed Bridgette finally made Bongo forget Colo.

Bongo and Bridgette were put together in April of 1985. By January 1986 a urine test confirmed that Bridgette was pregnant. The news of Bongo and Bridgette's impending birth arrived at about the same time as the tragic news of Dian Fossey's murder. She had been killed as she slept in her mountain cabin in Rwanda, while researching the mountain gorillas of Africa. Seven months before Bridgette was due to give birth, the keeper staff decided they would honor the primatologist's memory by naming the offspring—whether male or female—Fossey.

At the age of twenty-nine, Bongo was about to become the oldest sire in captivity. It had been nearly fifteen years since his last offspring, Toni, was born. This time around, Bongo would be given the opportunity to participate in the birth and rearing of his offspring. Instead of pulling him from the cage of his mate, which had been the practice when Colo was pregnant, Bongo would be present for the birth and remain an active presence in the raising of his offspring. After several failed attempts to allow female gorillas to raise their own offspring, keepers had high hopes that Bridgette would be Columbus's first gorilla to do it successfully.

The zoo community, led by people like Aspinall and Warren Thomas, then the director of the Los Angeles Zoo, had become concerned that gorillas would lose their natural parenting instincts altogether if zoos didn't allow them to raise their own offspring. Thomas argued that zoos were creating an entire generation of human-imprinted gorillas, ignorant of this essential social function. Beginning in 1973, when the Endangered Species Act was passed in the U.S. Congress, no more gorillas could be taken from the wild. Thomas and many others feared that as the wild-captured females moved beyond childbearing age, there would be no gorillas left in zoos that "knew how" to be mothers. The gorillas brought from the wild had theoretically seen and experienced their own mothers' care. Most of the captive-born animals had been raised exclusively by humans.

During the spring and early summer of 1986, the Columbus keepers studied reports of gorillas' being mother-reared in captivity. Though it had been tried several times, for one reason or another a gorilla infant had never been raised by its own mother at the Columbus Zoo. When representatives from the zoo had visited Howlett's to look at the zoo's gorilla habitat, Aspinall had said to Hanna, "I can't understand why you have never had a female care for her young." The group came back with a renewed commitment to make mother-rearing work in Columbus. "We have to take more chances," Hanna said. Risk taking, and the resulting rewards, would be one of Hanna's many legacies in Columbus. The keepers, of course, tried to minimize the risks in every way possible. They wrote letters, talked to other keepers, and visited the handful of zoos that had successfully allowed their gorilla mothers to raise their young. They reviewed stacks of data and learned everything they could about gorilla mother-rearing.

In Bridgette, the keepers had a wild-born animal who was also an experienced mother and thus a great candidate for caring for her own child. The biggest difficulty was Bridgette's history of excessive bleeding during labor and birth. The keepers would have to monitor not only Fossey's health but Bridgette's as well. If Bridgette lost too much blood, she might not be able to raise the baby on her own. To prepare Bridgette for the birth, the keepers put her on a special diet. They thought her history of bleeding could be mitigated by a tea of red raspberry leaves and alfalfa. She was also given foods high in vitamin K. To encourage milk production, Bridgette was given fennel seeds, beginning about six weeks before her due date and continuing after the birth of her baby. A special tea was given to her daily to promote milk flow.

The keepers cite the use of herbal teas during Bridgette's pregnancy as another example of Hanna's strong leadership. Hanna believed in allowing the keepers—the experts—to make decisions where the gorillas were concerned, as long as the veterinary staff approved the program. The keepers say that without the willingness and trust of Hanna and the veterinarians, Bridgette and her baby would have been put at greater risk when she gave birth.

On August 13, 1986, Bongo sat in the cage with his mate, quietly and calmly out of the way, while Bridgette gave birth to a male gorilla. Luckily, Bridgette did not suffer from abnormal blood loss. Shortly after Fossey's birth, Bridgette cleaned him off and inspected him carefully, seeming to look over each finger and toe. Her joy was evident as she looked down at her newborn.

Within twenty-four hours of Fossey's birth, Bongo was gently stroking his son. It wasn't easy to get access to the baby—Bridgette was very protective. Although Bongo approached the situation quietly, looking away as he reached out to the baby, Bridgette would forcefully push his finger away. In his frustration, Bongo would chew on his fingers, waiting for another opportunity to touch the baby.

Fossey had been born to what amounted to geriatric parents. Both Bridgette and Bongo were approaching the age of thirty. In the not too distant past, many zoos had given up on females' giving birth after the age of twenty, and males were thought too elderly if they had reached the age of thirty. Not only had Columbus successfully bred two valuable animals but the zoo had also made a strong point to other zoos around the country. Given the right circumstances, gorillas would breed well into their twenties.

The keepers could see how tired the baby was making the adults, but, like most new parents, they seemed to persevere as best they could. Bongo was an involved and attentive father. Whenever Bridgette got up in the night to feed Fossey, Bongo was right there by her side. Bridgette was a savvy and

Bongo cradles Fossey while Bridgette looks on

proud mother, happily sitting in plain view of the keepers as
she fed the baby, as if to communicate that "Everything is
fine." Just as the keepers had hoped, the Columbus Zoo had
its first mother-reared gorilla.

On Labor Day weekend, the new family made its public
debut. As the doors to the exterior habitat opened, Bongo led
Bridgette and Fossey up the chute and outside, where visitors
encircled the habitat, stacked five and ten deep, craning their
necks to see the baby. Once Bridgette figured out how to
make her way around the outdoor habitat while holding the
baby in one limb and using the other three to maneuver, she
hobbled around the perimeter of the circular play yard, dis-
playing her son to everyone in attendance. Bongo put on a
couple of impressive displays, pounding his chest and charg-
ing the mesh that separated the animals from the public.
Frisch says his actions were typical of a silverback protecting
his offspring. Bridgette made the keepers and the public ner-

vous as she acrobatically climbed the mesh high above the ground inside the enclosure. She held on to Fossey as she climbed but was forced to let go of him on a couple of occasions, leaving it up to the baby to cling to her hair.

Bongo and Fossey were responsible for another "first" for Columbus and its keepers. Once as the baby lay sleeping in his nest inside the family's cage, Bridgette went to an adjacent outdoor cage to forage. A work crew was in the gorilla house doing repair work and making a typical amount of construction noise. Keepers had discussed with them the presence of the baby and the importance of keeping as quiet as possible for his sake, but despite their best efforts, the workers' noise drowned out the sounds of an awakening Fossey as Bridgette sat outside. When Fossey stirred in the nest, Bongo started looking for Bridgette to emerge from the chute that led to the outside cage about thirty feet away. The baby's stirs soon turned to cries, but Bridgette remained outside. Keepers watched as Bongo seemed to consider his options, then decided to scoop up his son, complete with an armful of hay, in an apparent effort to comfort him. Frisch says Bongo looked uneasy when he had the baby in his arms. She told the workers to stop. Bongo walked upright around the cage, nervously holding the baby. Suddenly, Bridgette realized Fossey was awake and rushed back to the cage. She vocalized to Bongo and retrieved the baby. Frisch says, to her knowledge it was the first time a silverback in captivity had ever held an infant.

As the weeks passed, the keepers watched the family with great interest, and great joy. The gorilla staff and nursery staff observed as much of Bridgette's behavior as they could, hoping to learn the "gorilla way" of rearing an infant. Among other things, the keepers picked up on her comforting vocalizations to the infant, saw how she got down to his level to teach him to eat solid food, and watched as she taught him to ride on her back. When Bridgette wanted to rest, she would place her outstretched hand over her son's ankle. Her substantial arm acted as a leash, allowing her to lie down for a few minutes while Fossey moved about. The lessons the keepers learned from Bridgette have been applied with each subsequent gorilla baby that has had to be raised in the Columbus nursery.

Bridgette, Bongo, and Fossey were a healthy trio. Gradually, Bridgette allowed Fossey and his father to play together more and more. And Bridgette and Bongo's relationship was as strong as ever. Armstrong remembers Bongo sitting on the floor of his cage one afternoon, tossing around a ball. Bridgette was sitting across from him in a tire. The tire was squashed and filled with Bridgette's rotund body. As Fossey slept, Bridgette was watching Bongo toss the ball. At times, she would reach for it, as if she were trying to knock it away. Nearly all of the keepers were watching as Bongo put on his usual show of throwing the ball in the air and catching it without looking. On this particular occasion, Bongo seemed to be overdoing it for the sake of his audience, and he had all the keepers in stitches. Suddenly, he stopped tossing the ball and just looked at it. Then he looked at Bridgette. Then he looked at the keepers. Then he gently bounced the ball off Bridgette's head. She playfully flailed her arms and fell backward, as the two gorillas vocalized with what amounts to the sounds of gorilla laughter and contentment.

The keepers used to put a grain tub—the kind the zoo uses to feed hoof stock—in with the family. It could be used as a toy, or Bridgette would use it like a bassinet, filled with hay for Fossey to sleep in. One day, as Fossey was playing, Brid-

Bongo, Bridgette, and Fossey

gette decided to sit in the tub. Bongo began playfully tickling and wrestling with her. As she got up to get away from him, the tub stuck to her backside. Bongo and Bridgette vocalized together as she struggled to remove the tub from her body. Armstrong says these are just two examples of how much Bongo and Bridgette seemed to enjoy one another's company as natural friends and mates. Bridgette usually controlled the relationship, and Bongo showed her respect, unlike some of his treatment of Colo.

Bridgette got sick shortly after Fossey turned one year old. She nearly stopped eating. The zoo veterinarian decided to have her sedated and given a thorough physical. Following the examination, Bridgette and Bongo were separated so that Bridgette could rest. The door between their two cages was cracked enough for the baby to move back and forth between his parents. At times, Frisch observed Bridgette seeming to push Fossey toward the door, as if encouraging him to go to Bongo. She had also weaned him much earlier than normal, as if she knew she would be unable to provide for him much longer.

Bridgette died on October 7, 1987. Fossey was not quite fourteen months old at the time of his mother's death. Her death seemed to affect Bongo and Fossey deeply. Bongo sat by the door as Bridgette's body was taken out, constantly calling to her. Armstrong later wrote in the *Gorilla Gazette,* a collection of articles written by gorilla keepers worldwide and compiled and published four times a year by the gorilla staff at the Columbus Zoo: "Having watched Bongo go through an especially harsh mourning period after [Bridgette's] death, I realize now that my worst nightmare would have been to have had something happen to Fossey and having to see his father go through another devastating loss."

The Columbus keepers called gorilla keepers at several zoos to discuss the situation. Although some keepers from other zoos thought it unlikely that Bongo could act as Fossey's sole provider, Armstrong says there was never any question among the Columbus staff about what to do. There was no precedent for a silverback's providing total care for what amounted to an infant gorilla. But Columbus keepers say that they let the animals dictate most of their decisions, and Bongo made this decision pretty simple. They had faith that he would not neglect his son. So, under twenty-four-hour observation, Fossey and Bongo were left together. Meanwhile, the keepers began looking for a female to introduce to the two.

Bongo was known to be an attentive father, but he quickly convinced everyone that he was also an excellent provider. As the observers looked on each night, Bongo would brush all the hay together and make Fossey a bed right next to him, just as Bridgette used to do—even though Bongo rarely made a nest for himself. Even after hay was made available to him, Bongo chose to sleep on the bare floor, just as he had for his first twenty-five years of captivity. As a precaution, the keepers decided to feed the animals in separate cages, figuring Bongo, like all silverbacks, would hoard the food. The opposite occurred. Keepers were worried that Bongo wasn't eating enough because he always saved a large portion of his own food for Fossey. Convinced that he was a more than capable provider, keepers ended the twenty-four-hour observation after just three days.

The two played together constantly. Bongo, now past the age of thirty, seemed young again. He was magnificent in his tolerance, as his rambunctious and tireless son pulled, tugged, and played with him. Bongo would place his huge mouth over Fossey and tickle his belly as a human mother does with her child. Once again, Bongo was making the deep vocalizations of gorilla contentment. Bongo also did a fine job of handling Fossey's first crisis. When his son got sick, Bongo spent two days calming Fossey with soothing vocalizations, and holding him as he shivered with cold chills, until the virus broke its grip.

In time, Fossey's physical movements began to mirror those of his father. He took on the same stance and the same walk. Bongo never carried Fossey like a mother would, but the two were inseparable. The local media embraced Bongo's story, calling him the father of the year. Even the tabloid newspaper the *National Enquirer* published a two-page story on Bongo and son. By all accounts, Bongo was an exceptional

Bongo and Fossey playing in the gorilla villa

father. But Armstrong chooses to believe that he was not really an exception. She thinks most adult male gorillas, if given the opportunity, would be good providers.

After about a year a female was found to share life with Bongo and Fossey. Molly, a thirteen year old on breeding loan from the Kansas City Zoo, took to Fossey immediately. Frisch witnessed a classic motherly episode just hours after Molly and the males were put together: Molly looked Fossey over and began washing his ears.

Bongo's success was inspiring. His story helped bring the world's gorilla keeper community together. The efforts of the staff in Columbus to seek out the opinions of their peers at other zoos led to the founding of the *Gorilla Gazette*. Their networking also inspired the first gorilla workshop, held in Columbus in June of 1990. The workshop provided gorilla keepers with their first formal opportunity to exchange ideas on gorilla care. They shared stories and research regarding gorilla behavior, diet, births, and deaths. Bongo's likeness graced the logo of the first gorilla workshop. The workshops now occur regularly, and each one is hosted by a different zoo.

Sadly, just three months after the first workshop, Bongo died of a heart attack. Armstrong found him as she made her morning rounds. Her first stop when she arrived at the zoo was always at Bongo and Fossey's cage. Armstrong says his death had apparently come swiftly. When she found him slumped over in his cage, Molly was standing guard over his body, while Fossey was bent over, trying to look into his father's unseeing eyes.

Bongo's gentleness lives on in his son. Keepers believe Fossey learned from his experience with his father and is excellent with young gorillas. He seemed to befriend and protect youngsters Kebi and Jumoke when they joined his group in Columbus. He also takes to female gorillas very easily—most likely because of the care he received from Bridgette. It seems that Fossey, who is now on loan to the zoo in Little Rock, Arkansas, has acquired the best traits of his nurturing mother and gentle father. The keepers hope that when it comes time for Fossey to witness the birth of his own offspring, the lessons he learned from his parents will be passed on to another generation of gorillas.

5

The Second Generation: Emmy, Oscar, and Toni

In the late 1970s, the future of the Columbus Zoo gorilla family seemed to depend on the second generation of captive-born animals, the offspring of Bongo and Colo. By the summer of 1978, Emmy, the oldest, had reached breeding age, and Oscar wasn't far behind. Females reach breeding maturity at an earlier age than males. The keepers were hoping that Emmy and Oscar would breed and further expand Columbus's gorilla troop. Although there was concern about the brother-sister match, available research indicated that family in-breeding occasionally occurred among wild gorillas. The staff would have preferred not to breed Oscar with Emmy, but the Columbus Zoo was in the unique position of having six gorillas (the males Mac, Bongo, and Oscar, and the females Colo, Emmy, and Toni) who, with the exception of Bongo and Colo, were all directly related. Opinions differed,

Emmy and Oscar

but the prevailing view was that if the breeding of siblings would lead to the furthering of the endangered species, it should take place. This was a decision that would not be considered today but was dictated by the circumstances of the time, when gorilla breeding loans were rare. (The American Zoological Association's Gorilla Species Survival Plan was adopted in 1982 to facilitate animal exchanges that would result in suitable genetic diversity among the captive gorilla population. The SSP master plan was implemented in 1988.)

By 1977 Emmy and Oscar had been together for more than five years and seemed happy and compatible. Meanwhile, Toni, the youngest of Bongo and Colo's three offspring, was also approaching breeding age. Toni had arrived unexpectedly, just three days after Christmas in 1971, and she, like Oscar and Emmy, had been given an "award" name. According to her keepers, Toni was spoiled, like the youngest in many human families. Moreover, space constraints had necessitated her staying in the nursery years longer than she should have. Because she was physically small, her keepers were afraid to put her in the same cage with her brother and sister, and, furthermore, they wished to keep her separated from the two older gorillas, who they were hoping would breed. Toni was born a gorilla but raised like a human, as her mother had been.

At the age of five, Toni was finally moved from the nursery to a converted cage next to her siblings. Keeper Dianna Frisch walked Toni to her cage, holding her hand as they went. "It was as if we were saying to her, 'Okay, Toni. You're a gorilla again,'" Frisch says. Over the next eighteen months, Frisch observed several playful exchanges between Oscar and Toni. Since Oscar had not successfully bred with Emmy, Frisch began to think that maybe Oscar and Toni might have more success. Frisch says her suggestion of putting them together was met with great apprehension. Again, the idea of an inbreeding arrangement was controversial, and Toni was much smaller than Oscar, which led to a fear that she might get injured, even during normal play. Nonetheless, Frisch continued to fight to put them together. In the early summer, the perfect opportunity seemed to present itself when the animals were anesthetized for routine physical and tuberculosis examinations. Frisch pleaded with zoo officials to be allowed to leave open the door between Oscar's and Toni's cages so that when the two awoke, they could be together. Her colleagues agreed.

As the days and weeks passed, everyone was encouraged

by Toni and Oscar's ability to play together without getting too rough. Within weeks, the two were observed breeding. It was believed that, at the age of six, Toni was probably still infertile. It's rare for a gorilla to become pregnant at such a young age, and as the months passed, nobody suspected or expected Toni to be pregnant. In the spring of 1979, however, seven-year-old Toni was not eating well—although she was gaining weight—and there was concern that she might be ill. On April 25 Frisch was on her normal rounds when she chanced to arrive just as Toni began giving birth. Toni and Oscar had been separated for feedings, so she was alone in her cage. Upon giving birth, Toni looked down at her newborn and let out a frightened vocalization. She grabbed the infant with both hands and tried to stuff it back inside her body. She seemed to remain in a state of nervous confusion for several hours.

Although she later held the infant, Toni was never observed trying to allow it to nurse. In keeping with the practice of the day, at the first sign of trouble the baby was pulled to be raised in the nursery.

Toni's healthy female infant, the world's first third-generation captive-born gorilla, was given the name Cora following another naming contest. The winner submitted a rhyme as her entry:

"C" is for Columbus, the city of my birth.
"O" is for Ohio, the best place on earth.
"R" means that it is rare for me to be born in captivity.
"A" is April, when I started my life as a daughter to Oscar and Toni, his wife.

Appropriately, the winner was a grandmother from Lancaster, Ohio, since the birth also made Colo the first gorilla grandmother in captivity.

Since Cora was the product of parents who were brother and sister, the zoo gave the newborn a complete genetic checkup. Although doctors found Cora to be completely healthy, she had a slight webbing between her toes—perhaps a sign of genetic abnormality. Now nearing the age of twenty, Cora has never given birth herself, although no direct link has been found between her possible sterility and her genetic lineage.

The same year Cora was born, Jack Hanna arrived in town to begin his tenure as director of the Columbus Zoo. Only thirty-two years old, Hanna was relatively unknown and untested as a zoo director, having come from running his own animal menagerie in Tennessee. But his enthusiastic personality was attractive to the zoo board, and they decided to take a chance on the young man.

One of Hanna's first decisions as zoo director was to create an outside yard for the zoo's gorilla collection. Showing the marketing savvy that would define his career in Columbus, Hanna immediately recognized that the gorillas were the key to the zoo's success. Just three years before, the zoo had failed to raise enough money to build a "dream display" for the gorillas, so Hanna knew he would have to be creative with the financing, design, and construction of a new gorilla habitat if he was going to be successful. He personally knocked on the doors of Columbus business leaders and managed to raise $52,000 for the habitat. The largest donor was John McConnell, the founder of Worthington Industries. Hanna also convinced the Columbus Board of Realtors to make a sizable donation. With that relatively small amount of money, Hanna and his staff converted the zoo's former elephant yard into Columbus's first outdoor gorilla habitat.

Oscar, Emmy, and Toni in Oscar's Yard, the gorillas' first outdoor habitat

Oscar's group was chosen to move to the new habitat. The zoo hoped to add at least one more female to the group—there was room for as many as three more gorillas. Colo, Bongo, and Mac remained in the old gorilla house. Oscar, Emmy, and Toni were reluctant to explore the outdoors at first, staying close to the concrete structure that housed their indoor cages. But within an hour, as the zoo staff and major donors looked on, the three animals were running around the open area, playing in the grass, and even splashing in the yard's small pond. Emmy, who had never seen a tree in her life, showed her gorilla instincts by climbing one of the trees in the yard. The branches of the tree had been trimmed to stubs to ensure that no animal could climb high enough to escape the moated habitat. In their new play yard, Toni and Oscar continued to mate. Their compatibility was obvious: keepers and the public could observe the pair sitting outside with their arms around each other. Oscar appeared more relaxed in his new environment, which soon became known as Oscar's Yard. Hanna's first project as director had been a smashing success.

The staff was now ready to take on additional animals and to expand the Columbus gorilla bloodline. Shortly after the new habitat opened, Joansie arrived from the Buffalo Zoo as Oscar's potential mate. The seventeen-year-old female had never given birth in Buffalo, but Oscar was a proven sire. As cameras from the National Geographic Society captured the drama on film—the society was preparing a TV documentary on gorillas, which aired nationally in April of 1981—Joansie was introduced to her new group and her new habitat. As Joansie explored the area alone, experiencing the outdoors for the first time, Frisch verbally coaxed her along. When Joansie appeared comfortable with her new surroundings, the Columbus gorillas were let outside to meet their visitor. Joansie was frightened at the sight of the three gorillas and tried to retreat inside, but there was nowhere to go. She sat nervously by the entrance to the indoor cages, as Oscar aggressively displayed his dominance, pounding his chest and rushing toward the stranger with his arms flailing. After a few tense moments, calm prevailed, and Joansie was accepted into the troop.

Within weeks, Oscar and Joansie were seen breeding. And

within a few months of Joansie's arrival, the Columbus gorilla staff announced that *all three* of Oscar's females—Toni,
Emmy, and Joansie—were thought to be pregnant. National
Geographic brought its film crew back to Columbus, and the
baby watch was on. Oscar's appparent success as a sire made
world headlines and stirred the curiosity of the zoo community. Three females, all pregnant at the same time, was unprecedented.

Although the veterinary staff eventually determined that
Emmy was not pregnant, hopes remained high in the spring of
1980 that Toni and Joansie would give birth. Emmy was sent
to Cleveland to complete the agreement between the Columbus, Cleveland, and Buffalo zoos that had brought Joansie to
Columbus. Frisch accompanied Emmy to help with her introduction to Cleveland's silverback Timmy. According to the
agreement, Timmy and Emmy's first offspring would belong
to Columbus, and their second would stay in Cleveland.
Ownership of subsequent newborns would continue to alter-
nate between the two zoos.

Meanwhile, Toni's April due date came and went, and the
keepers became concerned. By mid-June she was long over-
due and acting strangely. She was under twenty-four-hour
watch when a keeper saw her suffer a violent seizure. Zoo officials didn't make the connection immediately, but it was
the same type of seizure Warren Thomas had observed her
grandmother suffer just weeks before Colo was born. After
witnessing the seizure, the keeper summoned Hanna, who
made a midnight call to his wife's obstetrician, James "Nick"
Baird. Baird had been a snake collector as a kid and had
talked about animals with Hanna on the few occasions the
two had met. Hanna wanted to know if Baird could take Toni

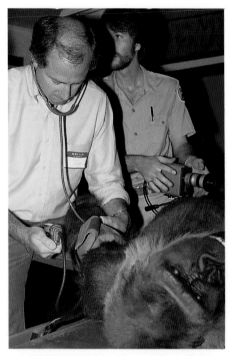

Dr. Nick Baird taking Baron Macombo's
blood pressure

to Columbus's Riverside Hospital, where Baird was on staff, and give her a complete examination. Hospital officials were reluctant to allow the gorilla to be brought on the premises but agreed to allow Baird to take whatever equipment and personnel he needed to the zoo. Hanna's call to Baird, and other experts like him, became another of his hallmarks. Throughout his tenure in Columbus, Hanna's friendly nature and aggressive pursuit of support from the community resulted in his calling on willing friends and acquaintances at all hours of the day and night. His call to Baird led to the doctor's lifetime of personal involvement with the zoo and its gorillas. In the mid 1990s, Baird became the president of the zoo's board of trustees.

When he arrived at the zoo to observe Toni, Baird's priority was to determine as quickly as possible if her baby was in any danger. Baird noticed that Toni occasionally postured as if she were having contractions. After noting several other symptoms, Baird diagnosed Toni with eclampsia. Although eclampsia had been thoroughly studied in human patients, it had never been reported in gorillas. Since the vets were unable to take Toni's blood pressure and confirm Baird's diagnosis, the doctor decided to call a colleague from the Ohio State University Hospitals to get a second opinion. Because eclampsia can complicate delivery, Baird was considering performing a caesarian section on Toni. Ohio State's Dr. Frederick Zuspan agreed with Baird's diagnosis but recommended a caesarian only if the animal continued to have seizures. Only one successful caesarian operation had ever been performed on a gorilla: in 1978, Los Angeles Zoo officials decided to deliver by C-section the fourth offspring of a gorilla named Ellie, who had previously killed three of her newborns.

Baird also thought the presence of the television cameras and crew from National Geographic might be complicating things by making Toni nervous. She seemed to stop her contractions when the cameras were present. His concern was apparently justified, because on July 26 Toni gave birth to a four-pound four-ounce male just after the camera crew took a break from filming. Baird and zoo officials were relieved, as the baby seemed normal and healthy. However, just ten hours after the uneventful birth, Toni seemed to lose interest in her baby. She began handling it carelessly, holding it upside down. Hanna, zoo veterinarian Harrison Gardner, and the veterinary staff made the decision to remove the infant from Toni and care for it in the nursery. "We felt we either had to take the baby away and have a nice healthy gorilla or wait for the mother to care for it and take the chance of losing the baby," Hanna told the press. When doctors reached the newborn male, he was cold and dehydrated. But with Baird's help, the infant, given the name Kahn, responded well to pediatric care.

Still wondering if his diagnosis of eclampsia was correct, Baird recovered the remains of the placenta and sent it to pathologists in San Diego. Tests revealed evidence consistent with maternal hypertension in human beings, supporting the diagnosis of eclampsia. It wasn't until Baird published his findings in the *American Journal of Obstetrics and Gynecology* that the connection was made between Toni's predelivery seizures and Millie's seizures two generations before. The cases prompted the zoo community to set up a database of gorilla births in captivity so that any abnormalities related to family medical history could be considered in future births.

A mere five days after Kahn's birth, Joansie surprised everyone by giving birth to a four-pound three-ounce male. Fewer than 150 captive gorilla births had occurred in history, and now two had taken place within a week in Columbus. Based on its agreement with the Buffalo Zoo, Joansie's newborn belonged to Columbus. The zoo was delighted to have the beginnings of a new gorilla bloodline.

Joansie, a wild-born animal, also failed to adequately nourish her infant. He was removed from his mother and taken to the nursery to join his half-brother, Kahn. Hanna was concerned enough about his gorilla mothers' lack of nurturing skill to consider using gynecologists' films in an effort to teach them how to nurse. Other zoos had used nursing films—and even explicit sexual films—hoping to teach their gorillas how to breed and to care for offspring. Hanna's public musings were picked up by the national media. Radio commentator Paul Harvey had fun with Hanna's ideas on his nationally syndicated show. In the days following the blitz of media coverage, Hanna received more than twenty calls from new mothers making serious offers to demonstrate breastfeeding to the animals. It seemed the mothers were appalled that the zoo would consider showing the gorillas films when live demonstrations were available. Hanna gratefully agreed and made arrangements to bring one of the anonymous volunteers to the zoo. Following the private demonstration, Hanna reported that Toni and Joansie showed some initial interest in the mother and her nursing baby, but Oscar dominated the scene. He pushed his mates aside and stuck his nose against the glass that separated him from the nursing woman. "Oscar watched the most," Hanna told the press. "He didn't seem to want to allow Toni and Joansie to see. We had to entice him with bananas to get him away so the girls could see." Hanna thought the demonstration was successful enough to plan for a series of similar demonstrations the following spring, when with luck Toni and Joansie would be ready to give birth again.

Hanna and the now-famous Oscar had thrust the Columbus gorilla program into the national limelight again. However, disaster struck. Kahn contracted salmonella and his health rapidly deteriorated. At the age of eighteen days, he died. Just as suddenly, his half-brother also got sick, and keepers feared he too might have contracted salmonella. The

Oscar lording over "his" yard

second infant, named Roscoe, would in fact spend most of his life fighting the effects of the illness, succumbing just two days before his first birthday. (Roscoe's life and death are recounted in chapter 7).

Meanwhile, Oscar continued to sire additional offspring. At the beginning of 1981, Joansie and Toni were both pregnant again. Their pregnancies were confirmed during a mass gorilla examination performed in May. An assembly-line checkup system, featuring twenty doctors, dentists, and veterinarians, drew a large crowd of onlookers and media. Each gorilla was tested for tuberculosis and given electrocardiograms and a dental checkup. Ten men were needed to lift Bongo onto an examination cart. Dentist Harry Barr used a two-by-four to wedge Bongo's mouth open during his examination. Once, while working on Bongo, Barr heard the board beginning to crack. Bongo was coming out of his anesthesia. The team of doctors quickly wrapped up their work and got Bongo back in his cage. On another occasion, Colo reached up and slapped Barr while he was drilling her teeth. While it seems humorous in retrospect, the caregivers treated the situations seriously. "Once, after transporting Colo off-site for some treatment, she came around in the zoo station wagon as we returned to the zoo," veterinarian Gardner says. "I looked in the back, and there was Colo, sitting up, looking out the back window." Gardner had to administer another quick shot of anesthesia to assure that she could safely be transported back to her cage at the zoo.

Four gynecologists examined Toni and Joansie. Joansie was given a due date between August 15 and September 1; Toni was thought to be due in October. The doctors used an ultrasound to view the gorilla fetuses and determine due dates.

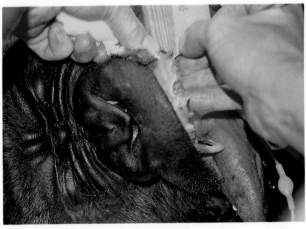

A dental exam

On July 31—two weeks early—Joansie gave birth to a healthy male. According to the breeding loan agreement, the infant belonged to the Buffalo Zoo. The baby was pulled and nursery reared, at the request of the Buffalo Zoo. When he was old enough, he was sent to Buffalo, where he lives today. He was given the name O.J., in honor of his parents, Oscar and Joansie, then later renamed Rich.

On September 13 Toni gave birth to her third and final baby with Oscar. Since the baby was a girl, she was to be sent, by prior arrangement, to the San Francisco Zoo. Columbus and San Francisco had worked out a deal that would eventually bring another breeding male from San Francisco to Columbus in exchange for the infant female. When the baby, given the name Zura, was pulled from Toni, doctors discovered that she had a heart defect—a hole between the two chambers of her heart, causing dilution of the oxygen in her blood. She was said to be in critical condition. However, after six weeks in the zoo hospital, Zura weighed more than four pounds and was showing improvement. Doctors hoped her heart was healing itself. She was taken to Ohio State University for an ultrasound and received treatment and medication. Zura eventually recovered fully and has lived a healthy life in San Francisco.

Although they continued to defend their decision to mate Oscar and Toni, the keepers decided that the brother and sister's three offspring were enough. Oscar had taken to Joansie, and the zoo was working on bringing another female to Columbus to join them. At the same time, artificial insemination was becoming more common in the zoological community, and researchers hoped it might be a reasonably priced answer to genetic diversification. Physically removing gorillas from their home zoos and shipping them around the country for breeding purposes was difficult for all concerned. Although just one gorilla (in Europe) had been successfully inseminated, and her infant had died within hours of its birth, researchers at the Yerkes Primate Center in Atlanta were confident that the process could work. Yerkes had successfully inseminated chimpanzees, and the center wanted to try inseminating Toni, a proven gorilla mother at a progressive zoo.

In October 1982 the keepers isolated Toni in the old zoo hospital. Frisch termed her living conditions there "deplorable" but hoped at the time they would be only temporary. When Toni ovulated in November, Ken Gould of Yerkes inseminated her with frozen sperm collected from Samson, a silverback from the Milwaukee Zoo. However, by the first of the year, it was clear that Toni wasn't pregnant. Attributing their failure to apparently nonmotile sperm, the researchers were still convinced that Toni was a good candidate for insemination. This explanation for Toni's failure to conceive was later disproved when Gould collected and tested sperm from Mac and Bongo. Neither of those specimens was very motile, either. Since Bongo subsequently became a father again, nonmotility came to be considered a characteristic of the gorilla sperm.

In April Gould tried to inseminate Toni again—this time with unfrozen sperm from a male at Yerkes. Again, the insemination was unsuccessful. Toni spent a total of eighteen months in "solitary confinement" in the poor conditions of the old hospital. Columbus considered sending her to another zoo because she had proven to be a very fertile animal and it didn't make sense to keep her alone. But the promise of the arrival of a breeding male from San Francisco (Sunshine eventually arrived in February 1985), and talk of building a new and larger gorilla facility in Columbus—the "gorilla villa"—kept Toni at home.

Toni was and is an unusual animal, with distinctive mannerisms, including a proclivity for walking upright like a human. She can also be seen sitting in her habitat, flapping her hands back and forth near her temples, as if she has a nervous tick. Beth Armstrong says she has her father's muzzle and her mother's intelligence. She attributes Toni's unusual behavioral habits to her human upbringing, her early and perhaps frightening experience as a mother, and her time alone in the old hospital. Frisch agrees that the unique circumstances in Toni's life have shaped a one-of-a-kind gorilla. Another one of Toni's trademarks is her attraction to jewelry. Baird says she is always fascinated with his watch and has offered to trade him some lettuce for it. Armstrong simply says, "Toni is an enigma. She's in her own world. It's 'Toni's World.'" She shows her intelligence by using some limited sign language the keepers taught her when she was in the nursery. When she's hungry, she gestures to her wrist, and then the palm of her hand, signifying that it is "time to eat." When Dian Fossey was at the Columbus Zoo, she visited Toni. Whereas Bongo reacted positively to Fossey's vocalizations, Toni responded to her overtures by reaching between the bars of her cage, grabbing Fossey's notebook, and tearing it up.

Toni's bipedal walking style intrigued Frisch so much that she had her knees X-rayed during one of her physical examinations. Unlike humans, gorillas don't have the locking mechanism in the knee that allows them to stand upright. The X-rays were inconclusive, but Toni continues to confound keepers and visitors by running across her exhibit on two legs.

A pregnant Toni standing erect

Frisch has even seen her walk through the opening of a door on two legs and duck as she entered, just like a human would.

As the zoo contemplated Toni's future, it received bad news about her thirteen-year-old sister. Emmy, who had been sent to the Cleveland Zoo, died unexpectedly in June of 1982. Keepers there said that Emmy had eaten a normal breakfast of bananas, carrots, oranges, and sweet potatoes, but when they checked her at 1:00, she was lying down in her cage, apparently very sick. By midnight, she was dead. Her body was

returned to Columbus for an autopsy amid complaints from
Columbus gorilla lovers. The zoo received dozens of phone
calls from zoo members who were angry that Emmy had died
at a "foreign zoo." Hanna was quick to state publicly that
"no one's at fault. No one is liable." Both Columbus newspapers ran editorials, eulogizing Emmy and mourning her loss,
but reminding readers that two Columbus gorillas were pregnant and, despite the loss of Emmy, the Columbus gorilla
breeding program was continuing to progress. The cause of
Emmy's death was determined to be peritonitis.

Meanwhile, Columbus secured another female to take
Toni's place with Oscar. Bridgette had arrived in Columbus
in 1981, on loan from the Henry Doorly Zoo in Omaha, Nebraska. Bridgette (whose later mating with Bongo was described in chapter 4) was wild-born and a proven mother. She
would provide another important bloodline for the Columbus gorilla population. It didn't take long for Bridgette and
Oscar to mate. By the summer of 1982, both Bridgette and
Joansie were thought to be pregnant. Unfortunately, Bridgette miscarried in the fall. Joansie's pregnancy proceeded
normally, and she was expected to give birth early in 1983.

During a New Year's Eve party at the Columbus Athletic
Club, Hanna received the call that Joansie had given birth.
The infant had been born just five minutes before the stroke
of midnight. Another partygoer, overhearing Hanna's end of
the conversation and mistakenly assuming he was receiving
the news that his wife had given birth, congratulated him and
asked where the baby was staying. "Oh, he's still with the gorilla," Hanna said. Hanna remembers the stranger, who evi-
dently didn't realize he was talking to the Columbus Zoo
director, responding to his statement with a look of confu-

sion. "I told him it was too complicated to explain," Hanna said. Joansie's male offspring, given the name Lang, was the twelfth gorilla to be born in Columbus, and he made Oscar a father for the sixth time.

Oscar's life as a prolific breeding male continued in 1983. Bridgette became pregnant, and in July the zoo made the stunning announcement that an ultrasound had shown that she was carrying twins. They would be the the first gorilla twins ever born in the Western Hemisphere. Amid an international media blitz rivaling the one that surrounded the birth of Colo, Bridgette delivered a pair of healthy males in October. (Their story is described in chapter 6.)

In August of 1984, two months after the death of the family patriarch, Baron Macombo, at the age of thirty-eight, nineteen-year-old Mumbah, a silverback from Aspinall's Howlett's Zoo, arrived to lead Columbus's other gorilla troop. Hanna and Columbus Parks and Recreation director Mel Dodge went to Detroit to meet the British Airways flight carrying Mumbah from Britain and drove the sedated animal to Columbus. The local media called it a "British invasion" and decided Mumbah was being shipped to Columbus to "teach our gorillas how to swing" in the zoo's Aspinall-inspired "gorilla villa," which had opened just three months before Mumbah's arrival. Although Mumbah had sired one offspring at Howlett's, he has yet to breed successfully with any females in Columbus. However, and perhaps more important, his agreeable nature as a silverback has been key in allowing Columbus to build a uniquely age-diversified group of gorillas.

The following February, 1985, Sunshine, a silverback-to-be, arrived from San Francisco. He would lead a third

Sunshine, at 6'5" one of the world's tallest
gorillas

Columbus troop, which would include Toni. Toni and
Mumbah's lack of breeding success led keepers to introduce
her to the gangly eleven-year-old newcomer from California.
As Sunshine grew into perhaps the largest gorilla in captivity,
with a height of six feet five inches and a weight well over 500
pounds, he began to rival Oscar as a proficient sire. Between
1987 and 1991 Toni gave birth to four infants sired by Sunshine, three males and a female. Sunshine has also sired three
offspring with Lulu, and two offspring with Molly. Of Toni's
infants, only Norman, left with Toni in hopes that she would
raise him herself, did not survive. He died of malnourishment

at the age of ten days. The failed attempt to allow Toni to mother-rear an infant proved how difficult it can be to try to allow nature to take its course. Despite being given every opportunity to raise her own offspring, Toni has never been successful. No one will ever be sure why Toni reacts to her offspring the way she does. Nonetheless, she has contributed a great deal to the success of the Columbus gorilla program, giving birth seven times over a twelve-year span.

Oscar, meanwhile, also continued to breed successfully, siring another male infant with Bridgette in 1985, and siring two females and a male with Pongi, a wild-born female sent to Columbus on breeding loan from the Birmingham Zoo in Alabama. Pongi and Oscar's first baby was Oscar's tenth. For the first time, Oscar was given the opportunity to be present at the birth of one of his offspring. As Pongi gave birth to Mwelu, he sat quietly about ten feet away. Apparently realizing the baby was all right, he moved to the corner of the cage and sat motionless as he watched Pongi attend to the newborn. His keepers say Oscar was a tolerant and attentive father, once he was given the opportunity to interact with this offspring. He seemed to especially enjoy Pongi's second-born, Colbridge, a male whose name is a combination of the words Columbus and Bridgette.

Oscar was responsible for perhaps Columbus's closest call with a gorilla escape. While about 1,500 people were enjoying a summer evening at the zoo in 1982, Oscar was discovered wandering the keeper aisles outside his cage. Hanna was in his office, on the telephone, when the emergency call came. "We've got a gorilla loose!" Hanna shouted as he ran the four hundred yards across the park to Oscar's Yard. By the time he got to the habitat, the area had been evacuated, and the Columbus police SWAT team was on the scene with their weapons drawn, ready to shoot to kill if Oscar escaped from the building. There was only one door to freedom for Oscar, and Hanna moved as fast as he could to have a construction crew, on the zoo grounds as part of a construction project, move a backhoe up against the door. Once Oscar heard the big piece of machinery, he began hitting or kicking the door, apparently trying to get out.

Just as he would do when Colo would escape to the keeper aisle of the main ape house a few years later, veterinarian Gardner climbed into the rafters above the animal cages. Armed with a rifle powered by carbon dioxide cartridges, Gardner listened to Oscar wreaking havoc and tried to get a clear shot at the animal. After several minutes, Oscar exposed an arm well enough for Gardner to take a shot at it. Unable to see how Oscar reacted to the dart, Gardner relied on his wristwatch to time the effects of the tranquilizer, and quietly entered the building carrying a jab syringe for safety. He found Oscar lying on the floor and, with the help of three men, dragged him back to his cage.

In June of 1993 Oscar was knocked down for his annual physical examination. He never awoke again. The veterinary staff determined that he had had a heart attack in his cage as he was coming out of the anesthesia. He was twenty-three years old. Just six weeks after his death, Pongi gave birth to Oscar's twelfth offspring, Casode, who was mother-reared in Mumbah's group. Oscar is remembered for his playful nature, his huge chest, and his resemblance to his mother, Colo. He will also be remembered for his contribution to the Columbus Zoo gorilla program, as well as many other gorilla programs across the United States who now house and breed animals sired by one of the most prolific captive gorillas in history.

6

The Twins, Mosuba and Macombo II

Oscar will perhaps be best remembered as the sire of the first gorilla twins born in the Western Hemisphere. The twins gained the attention and appreciation of a whole new generation of zoo supporters and gorilla lovers who have enabled the Columbus gorilla program to continue to prosper.

Following the discovery in July 1983 that Bridgette was carrying twins, the zoo mobilized more than sixty volunteers who worked around the clock to ensure that she was comfortable and that her babies would be born healthy. Bridgette had suffered miscarriages during her previous two pregnancies in Columbus, and the gorilla staff made it their mission to see that she carried the twins to full term. She was given special vitamins and comfort foods, such as sunflower seeds, that the keepers hoped would help her remain content. She was also given fresh hay each day for nest building. Oscar was separated from his mate to prevent any rough activity that could result in accidental injury to Bridgette or the unborn infants.

127

A "Bridgette Watch" was organized starting on September 1, 1983, six weeks before the due date established by the veterinary staff. Based on his knowledge of the complications associated with human twin births, Dr. Baird prepared the zoo for the likely possibility that Bridgette would deliver prematurely. And since human mothers who are carrying twins are more susceptible to hypertension, toxemia, and eclampsia, Baird and the zoo staff also watched Bridgette closely for signs of these conditions. As Bridgette entered her final trimester, Baird monitored the fetuses' position in the womb and was prepared to intervene during labor if necessary. Video cameras were mounted in the observation loft above Bridgette's cage along with two still cameras. A room in the research hospital was converted into a watch station where volunteers and zoo staff could monitor Bridgette's every movement without disturbing her. On October 25, around 10 P.M., while Jack Hanna's wife, Suzi, was on watch, Bridgette's water broke, and active labor began. Suzi had a few anxious moments as she nervously tried to start the video recorder and use the walkie-talkie to summon help. Bridgette was visibly uncomfortable when Dianna Frisch arrived, and looking at her from the public viewing aisle, Frisch could see that she was dilating. Shortly after midnight, Bridgette moved to her nest. With one long push, her first baby was born. She immediately picked up the infant and cradled it in her arms. Just fifty-seven seconds later, the second baby was born. As Bridgette turned to pick up the second newborn, her attention was diverted to the placenta, which she began to consume, a natural thing for a gorilla mother to do.

The excited and nervous observers were all focused on the second infant, who, like Colo, had been born while still in the

Baby B, still in its amniotic sac

amniotic sac. Dr. Baird alerted everyone that he and the vet staff only had a couple of minutes to break open the sac and allow the infant to breathe. As precious seconds passed, Bridgette's attention remained fixed on the first infant. The decision was made to move her to an adjoining cage so that doctors could get to the second baby. Oscar was sleeping peacefully in the cage where keepers wanted to move Bridgette, but as soon as he was aroused, he moved to an outdoor area to make room for Bridgette and the first twin. Lured away by bananas, Bridgette, with the first twin clinging to her chest, moved to Oscar's cage as doctors rushed in with a small hook and broke open the second twin's amniotic sac. The infant immediately let out a newborn cry and, to the relief of all, began breathing on its own. Baird wrapped it in a towel and headed out into the crisp autumn night, walking at a brisk pace toward a waiting incubator in the zoo hospital. With a bundle of history in his hands, Baird got lost in the dark, unfamiliar area of the zoo. "Imagine this scene: I took off for the nursery with this priceless newborn in my arms, and I got lost," Baird recalls. "I never got to the point where I was panicked, but it was not a comfortable situation. And it was cold as hell!" The doctor had gotten turned around in the zoo's maintenance area. Luckily a zoo employee, sent out to find him when he didn't arrive at the hospital, located the doctor and escorted him to the nursery.

The second twin was a healthy male, given the name Baby B. Meanwhile, the keepers gave Bridgette about forty-five minutes to calm down before giving her anesthesia in preparation for removing the first twin. Frisch got worried as she watched Bridgette get woozy. What if she accidentally fell on the baby? Before Bridgette was totally unconscious, Frisch quietly opened the cage door, reached in, and gently removed the infant from its mother's grasp. Baby A was also a male. As Frisch rushed him to a second awaiting incubator, veterinarian Gardner gave Bridgette a quick exam and obtained about 10 cc of colostrum for the infants. Colostrum is the breast milk secreted for the first few days after birth; it is particularly rich in protein and antibodies, which would help the twins build their natural immunity to disease and infection. Bridgette had lost a lot of blood during the delivery, but Dr. Gardner noted that the bleeding had all but stopped, and she appeared to be resting comfortably.

Because gorillas so rarely give birth to twins, the Columbus keepers did not have much knowledge to work from. Although there have been some reports of gorilla twinning in the wild, multiple births are more rare in gorillas than in humans. The latest known birth of gorilla twins in the wild was in 1986, to a female in mountain gorilla group 5, a set of animals being observed by primatologists in Rwanda. At the time of the Columbus twin births, only two pairs of twins had been born in captivity. The first pair, both female, had been born at the Frankfurt, Germany, zoo in 1967. Unlike the Columbus twins, who at first glance looked exactly alike, the Frankfurt twins were said to be distinct in both physical appearance and color and, as they grew, their behavior also became distinct. One of the Frankfurt twins died before she reached the age of two. In 1981 another set of gorilla twins, one male and one female, was born at the zoo in Barcelona, Spain. The male survived just seventeen days. The father of the twins was the famous all-white gorilla named Snowflake, although neither of the infants was born with their father's white coat. In the United States, Kribi Kate, a female at the Kansas City Zoo, aborted two fourteen-week-old fetuses in 1966; and in 1981 Oko, a female at the Yerkes Primate Center, gave birth to twins, but one of the animals was stillborn.

The Columbus keepers were concerned because in both of the earlier surviving sets of captive-born twins, one of the pair had died early. However, two positive factors were reassuring: Bridgette had carried the infants to full term, and their birth weight was substantial. In fact, at four pounds eight ounces apiece, they both weighed more than most single babies, and their lengths, measured at eleven and ten inches, were also on the high side for the average newborn gorilla.

By 2:00 A.M. the staff realized they had two new healthy gorillas, and the zoo staff and volunteers celebrated with champagne. The keepers suddenly realized they had witnessed one of the rarest events in zoo history. As they relived the previous hours, smiling and toasting their good fortune, Frisch remembers thinking about Oscar and feeling that he should be part of the celebration. She left the party, carrying

The newborn twins

an unfinished bottle of champagne in her hand, and walked over to see him. As she entered the gorilla house, she saw him sitting quietly, watching Bridgette recover. When Frisch began congratulating him, and telling him what a good job he had done, Oscar got up and walked over to her. She offered him a celebratory sip of champagne.

Although the births and the hours that followed had gone well, Hanna tempered his excitement when talking to the local media. He repeatedly warned that the first few days of newborn gorillas' lives are uncertain at best and that their condition should be described as "guarded." In truth, they were doing wonderfully. The only complication was a minor navel infection suffered by Baby A.

Although the twins' continued good health was the first priority, the breeding loan agreement between Bridgette's home zoo (the Henry Doorly Zoo in Omaha) and the Columbus Zoo posed an unresolved problem. When Bridgette arrived in Columbus, nobody imagined she would give birth to twins. The loan agreement, signed by both zoos, specified that Bridgette's first offspring would belong to the Doorly Zoo, her second to Columbus, and so on. However, when an ultrasound had first revealed the presence of twins, Hanna called Doorly Zoo director Lee Simmons to discuss the implications. Both men agreed that the first twin born would be the property of Omaha, and the second would belong to Columbus. That arrangement seemed fine in theory, but seeing the two healthy, active, identical infants lying side by side, no one could contemplate separating them. Columbus Parks and Recreation director Mel Dodge, who had been a vocal and important supporter of the zoo, stated emphatically in the newspaper, "We're not splitting up the twins." The statement was, in part, a reaction to public outcry in Columbus. When the media reported that the infants would probably grow up in different zoos, the zoo's telephones started ringing off the hook. "We're going to work something out with the Omaha Zoo," Dodge said. "People who operate zoos are gentlemen." Hanna was growing weary of receiving phone calls at all hours of the day and night from people pleading with him not to split up the twins. The calls weren't coming just from Columbus; people were calling from all over the country—especially from parents of twins. Hanna reported receiving one call from as far away as Santa Monica, California, from a woman who kindly lectured him on the importance of keeping twins together, whether they are humans or not.

Since the zoo in Omaha closed for the winter at the end of October and wasn't scheduled to reopen until the following April, the zoos had some time to negotiate. It was decided that both twins would remain in Columbus until at least the spring, when Omaha's doors opened to the public again. Hanna thought the five-month respite would give the two zoos plenty of time to reach a settlement. His first thought was to offer Omaha one of Columbus's young females, knowing that Omaha had a four-year-old male who would eventually need a mate. When Simmons came to Columbus to see the twins in person, he told the anxious press that he, too, was interested in seeing the twins stay together. Of course, he wanted to see them together in Omaha. "We'll settle down and come to a reasonable decision, agreeable to both institutions," he said. After a visit to the twins, Hanna took Simmons over to the gorilla house to introduce him to both Toni and Cora. Afterward, the two men discussed the possibility of a trade between the two zoos involving one of the females. Hanna hoped that if one of the Columbus females could produce an infant in Omaha, just as Omaha's Bridgette had produced infants in Columbus, both zoos would be happy. The two directors also discussed a plan that would keep the twins together in Columbus for a length of time, perhaps a year or two, after which they would move to Omaha for the same time period.

Dodge presented his own tongue-and-cheek solution, offering Omaha "a dozen snakes, fourteen rare birds, a first-round draft pick, and [Columbus Zoological Society president] Bill Wolfe." Wolfe responded in kind, saying, "I'm not going. Omaha is too darn hot in the summer and too cold in the winter." Joking aside, the situation was serious. Not only was the animals' health at stake, but a separation of the twins could be devastating to the zoo's bottom line. The twins had created quite a stir in Columbus. Six hundred people were lined up outside the gates before the zoo even opened on the day of their public debut. In all, two thousand people crowded around the window of the zoo nursery on that

Wednesday, a day when typically about four hundred people would visit. Nobody wanted to talk publicly about the gorillas' monetary value, but it was certainly a major consideration as the two zoos tried to strike a deal.

A week after Simmons's visit, he sent the Columbus Zoo a letter laying out three options: First, the oldest twin, Baby A, could be sent to Omaha, but Simmons acknowledged that this was the least favored option. The second option involved Columbus giving four-year-old Cora to Omaha in exchange for Omaha's rights to one of the twins. It would be a permanent swap, involving ownership of the animal, rather than a breeding loan. The final option involved Cora's being sent to Omaha on an extended breeding loan, with an understanding that *all* of her offspring would belong to Omaha. Columbus officials didn't like any of the options. As Toni and Oscar's firstborn, Cora represented the important fourth generation of Columbus gorillas. Giving her away on a permanent basis was out of the question. And even though she wouldn't reach sexual maturity for several years, a breeding loan that released the rights to all her offspring was also not acceptable.

In reply to Simmons's letter, Columbus offered Omaha Toni on breeding loan. The proven dam would likely give birth in Omaha within a year and provide the zoo with an infant. Once the infant was determined to be healthy, Toni would then return to Columbus and the zoos would call it even. Simmons would not accept Columbus's offer. He was much more content to have a "gorilla in the hand." Gambling on Toni's breeding successfully, even though she was a proven mother, was not worth the risk.

Both zoos tried to negotiate in good faith but could not come up with a permanent agreement. Instead, they reached temporary common ground by deciding that the twins should travel between the two zoos for the short term. They drew up a schedule that called for them to travel to Omaha on March 30, after five months in Columbus, just in time for the Doorly Zoo's opening day. The twins would remain in Omaha for two months, then they would be returned to Columbus in time for the Memorial Day weekend dedication of Columbus's new gorilla habitat, the Howlett's-inspired "gorilla villa." The twins would return to Omaha for short stays on two more occasions before the Doorly Zoo closed for the season, just after the twins' first birthday.

Moving the twins 815 miles from Columbus to Omaha would be a daunting task. They had received twenty-four-hour care from the moment of their birth, mostly from nursery workers Chriss Pendleton, Molly Widdis, Sue Allen, and Barb Jones. The four took turns working seven-hour shifts, recording the twins' temperature, food intake, weight gain or loss, sleep and waking habits, quantity and quality of stool, and any other pertinent information. In addition to keeping detailed records, the nursery workers diapered, fed, cuddled, and played with the twins. They were also responsible for housekeeping and laundry chores. Like human twins, the gorilla infants required a lot of care, and the nursery workers worked full shifts.

Some zoo visitors criticized the look of the twins and the zoo nursery, with its brightly colored walls, human toys, and other familiar human nursery accessories. The twins were usually dressed in diapers and a jumpsuit. Some feared the nursery workers were overly humanizing them, but the zoo defended its practices, saying the infants' social needs were important. As for the diapers, their use was simply a matter of practicality. Not only did they keep the nursery and the gorillas clean, the diapers also made it easier for the attendants to collect data on each animal's stools. Furthermore, the zoo explained that while the clothes might look cute, they served a practical purpose by keeping the infants warm. The zoo provided stimulation using colors and human toys, but what the public didn't see and hear were the wild gorilla sounds that were being piped into the nursery, or the leaves, twigs, and other natural objects mixed in with the toys.

As moving day approached, the nursery staff wondered how the gorillas would react. Except for their first few moments of life, they had never been outside the nursery. The twins' four caregivers, along with Dr. Gardner, would make the trip. They packed the zoo van for the fifteen-and-a-half hour journey like a family going on vacation, loading up diapers, baby formula and food, clothing, portable cribs, and favorite toys, including a plastic teddy bear and a baby duck. The keepers dressed the twins in shirts that read "Goodbye Columbus" on one side and "Hello Omaha" on the other.

Barb Jones says the trip was indeed like taking kids on vacation. The gorillas grew restless and even seemed to bicker with one another. They were given treats to placate them, including food from a fast-food restaurant's drive-through window. In the end, they survived the trip in good shape. On their arrival in Omaha, the Associated Press called them "two brown-eyed, black-haired bundles of zoological significance." Their presence in Omaha drew the biggest opening day crowd in that zoo's history, estimated at 7,700. (The previous year's opening day had drawn only 800.) The Columbus keepers made daily calls to Omaha to check on the twins' progress. The keepers in Omaha reported that the gorillas

Jack and Suzi Hanna with the twins

were eating well and learning to swing from the jungle gym in the Doorly Zoo nursery. By the time they were delivered back to Ohio for Memorial Day, the Columbus nursery staff had added some hanging ropes, a sliding board, and a spinning top to their already impressive toy collection.

At nearly eight months old, the twins now weighed fifteen pounds each, but they were still being called Baby A and Baby B. While they were in Omaha, Columbus Zoo docent Sue Allen had appeared on the network television game show *Family Feud* and had told the show's host, Richard Dawson, about the twins. Dawson had encouraged the show's viewers to send name suggestions to Columbus, and the zoo received hundreds of ideas from around the country. Mel Dodge had his own idea for naming the twins. He suggested to the zoo's board of trustees that the gorillas be named for the mayors of Columbus and Omaha. Baby A would be called Mike, in honor of Omaha mayor Michael Boyle, and Baby B would be named Buck, after Columbus mayor Dana "Buck" Rinehart. Although both mayors professed to be flattered by the suggestion, Simmons was not pleased. He responded angrily to the proposal by telling the *Columbus Dispatch,* "As far as I'm concerned, any name that Omaha did not pick is out! The agreement is reasonably clear. I indicated up front that we would prefer to name our gorilla or that they be named by whoever ends up with both of them." The controversy pushed the zoos to decide the question of the twins' names, and by mutual agreement Baby A was given the name Mosuba, in honor of the three volunteers—Molly, Sue, and Barb—who had taken such good care of the twins during their first months. Baby B was named Macombo II after Columbus's patriarch, Baron Macombo, who had died before the twins' first birthday.

The joint custody arrangement continued through the summer and fall of 1984, until the twins returned to Columbus for the winter months. On their first birthday, the McDonald's restaurant chain donated a 266-pound birthday cake, which was enjoyed by 1,800 zoo goers, mostly children, who had been invited to the birthday celebration. The birthday celebration was broadcast live on *Good Morning America*—as were the twins' subsequent three birthdays. Bolstered by the twins' presence, the zoo reported an increase in its annual attendance of more than 80,000 visitors, representing more than $350,000 in additional gate revenues.

By the spring of 1985 the twenty-month-old gorillas seemed used to the rigors of travel, and the zoo staff had also become accustomed to the routine. Since fewer staff members were needed for each trip, air travel, which would be faster and more efficient, became a possibility. Hanna telephoned Dale Eisenman, the president of Zee Medical Products, a Columbus-based medicine distributor. Hanna's and Eisenman's daughters were high school classmates and good friends. Six weeks earlier, Hanna had called on Eisenman when a major snow storm threatened to keep him from making it to New York City to appear on the *Late Night with David Letterman* show. Eisenman was a private pilot, and he agreed to fly Hanna and his animals to New York. Now Hanna wanted to know if Eisenman would take a couple of zoo staffers and the gorilla twins to Omaha in his six-seat, twin-engine Cessna 310. Eisenman remembers Hanna showing up at the airport on May 16, frustrated because the nervous gorillas had soiled the zoo van with diarrhea. The scared gorillas hugged their keepers tightly as they were prepared for their first flight. Although the keepers had brought two cages for the growing gorillas to fly in, the twins insisted on being crammed into just one of them so they could stay together. As the plane took off, Eisenman cringed as the animals continued to have loose bowel movements. He wondered what his partner would think when he saw what the gorillas had done to the company aircraft. At the end of the otherwise uneventful flight, Eisenman remembers the excitement growing as the plane approached the Omaha airport. Aware that a greeting party, including many members of the media, were awaiting their arrival, Eisenman announced to the tower, "This is 2-0-9 Whiskey-Bravo on final approach—*with the gorillas!*" (Eisenman later became a zoo trustee and a member of its executive committee, overseeing development and strategic planning.)

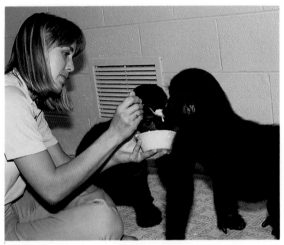

Keeper Beth Armstrong feeding the twins

For the first three years of their lives, the twins spent about four months a year in Omaha and eight in Columbus. Sally Boysen, a psychology professor at Ohio State University and an expert in great ape behavior, became concerned about the effects of the back-and-forth life the twins were leading. However, once she spent some time observing the twins during one of their prolonged stays in Columbus, she was surprised to find that not only had the two-city existence not been detrimental to the animals, it appeared to have enhanced their development by stimulating them mentally and physically. Boysen said Mosuba and Mac seemed healthier than most gorillas their age, and more playful and curious. Hanna and Frisch both concurred, saying that the twins were as well adjusted and healthy as any gorilla infants ever raised in Columbus.

But by the spring of 1987 it was obvious to the Columbus keepers that the twins needed to settle into a social group. In

The twins outdoors

July both zoos agreed to allow the twins to join Mumbah's group in Columbus. The three and a half year olds were introduced to the troop slowly. First, they met Cora. The twins held hands as they took turns approaching their half-sister. Once they were comfortable in Cora's presence, they were gradually introduced to the rest of the troop. All the while, the pair acted as one, making nearly all of their movements in unison. Colo immediately showed a maternal attachment to the twins, which helped facilitate their socialization process. Together, the twins were ornery and made life interesting for the keepers and the animals in their group.

A few weeks after the twins' introduction into Mumbah's group, a nursery-reared infant named J.J. was also added to the group as part of the zoo's surrogate program. Beth Armstrong says it was the first time the twins had seen a gorilla younger than themselves, and they treated him the way bigger kids in a neighborhood tease a newcomer. Even when J.J. approached the twins calmly and gently, they found a way to playfully torment him. However, if one twin felt the other was getting too rough, he would come to J.J.'s aid. Armstrong says this created a nice balance and seemed to keep the situation from getting out of hand. Moreover, if things got too rough in the eyes of their grandmother, Colo—who was J.J.'s surrogate mother—she intervened and seemed to send a clear message to the twins, through vocalizations and physical punishment, that their actions were inappropriate.

Despite the twins' natural adjustment to life in Mumbah's group, the keepers knew it was only a matter of time until they would have to be separated. Most available research indicated that in the wild the two would eventually go their

Colo and the twins, her grandsons, in the gorilla villa

separate ways, searching for their own troops to lead as silverbacks. Armstrong, however, believed that Mosuba and Mac represented an excellent opportunity to see if two males could lead a group together. Having watched them develop for nearly seven years, she was sure that if any two males could colead a troop, the twins could. She argued that there was a surplus of males in captivity, therefore the zoo community would not suffer if the twins were left together.

Nonetheless, the decision was made to separate them. In August of 1990, though they had never spent even one night apart, Mosuba was pulled, anesthetized, and sent to Omaha. Mac seemed confused by the separation. He spent his first seven days without his brother in virtual solitude, in self-imposed isolation from the rest of the group. Although he ate normally, he seemed distant. When he did begin to mingle with the other members of the group, he was more aggressive with J.J., which put the smaller animal at a disadvantage. Soon the adult females began to intervene on J.J.'s behalf. A positive result of their intervention and discipline was a heightened level of interaction between the juveniles and the adults. According to the keepers, Mac was beginning to be disciplined by the adults more like a blackback, instead of a juvenile. Two weeks after Mosuba's departure, the group dynamics in Mumbah's group had changed, but the family unit remained stable.

The group's dynamics changed once again a few weeks later, following Bongo's unexpected death. Bongo's two female mates and his son Fossey were slowly introduced into Mumbah's group. The keepers say that Mac's new blackback status was apparent as he protected the younger Fossey from the aggression of other members of his group. After watching him closely, they concluded that he had not been adversely affected by the separation from his brother; rather, he seemed to have matured as a result of the experience. The Doorly Zoo reported that it had to field questions from the public about Mosuba's mental state. Visitors were concerned that the separation from his brother had made him depressed. Those who had seen the rambunctious twins together were convinced that Mosuba was acting differently in the absence of his brother. It is true that on the day of his arrival Mosuba was reluctant to join the other gorillas in the zoo's enclosure, choosing instead to cling to his human attendants, leaning against the display glass where they were standing. According to the keepers, he would even choose the company of an unfamiliar human over another gorilla. However, on his second day in Omaha, he was introduced to his five-year-old brother Motuba, who had been born to Bridgette and Oscar in Columbus. It was as if he'd found Mac again. The two youngsters chased each other, wrestled, and played themselves into exhaustion.

In both Columbus and Omaha, the keepers seemed satisfied that they had done the right thing. Although there were no data on the effects of separating twin gorillas, both Mac and Mosuba seemed to adjust quickly to their new situations. Keepers at both zoos say they were attentive to the physical and social health of the animals during the difficult separation period, and the end result was positive for the two institutions.

Mosuba became a father on October 9, 1995, when his sperm was used to sire the first test-tube gorilla infant, Timu, born at the Cincinnati Zoo. Mac is a maturing blackback, with the mischievous personality of a human teenager, and remains in Mumbah's group in Columbus. Soon he will reach sexual maturity and will have to be separated to form his own group. Some of the keepers feel Mac may represent the future of the surrogate program. Mumbah's advancing age means the keepers must identify another appropriate male who, as a silverback, has the temperament to accept unfamiliar animals into his group. Mac's experience with Mumbah and as part of the current surrogate group makes him an obvious choice to fill that role when he reaches silverback status. In the meantime, he is a popular and strong animal who seems to enjoy the presence of visitors. He often plays to the crowd, much to the delight of Columbus gorilla lovers. Just as thousands of people flocked to the glass outside the zoo nursery to watch him tumble and play with his twin brother, thousands continue to visit the zoo to see the silverback-to-be playfully interacting with the members of his group.

7

Roscoe

The good health of the gorilla twins, Mac and Mosuba, and the confidence of the zoo veterinary and nursery staff in providing their care, can be attributed in part to the experience gained several years before during the care of another Columbus infant gorilla. Roscoe was the first offspring of Oscar and Joansie, born unexpectedly on July 31, 1980, just five days after Toni had given birth to Kahn. Roscoe was found as a newborn clinging to Joansie during the keepers' routine morning checks. The Buffalo Zoo had been so sure that seventeen-year-old Joansie would not breed that they agreed to allow Columbus to have her first born, an unusual concession for a zoo sending a female to another zoo on breeding loan.

Although Kahn had had to be pulled from Toni, the zoo had decided to leave Roscoe with Joansie, because she appeared to be taking excellent care of him. However, within a week's time it became apparent that Joansie could not adequately nourish her infant. As keepers kept an around-the-clock watch on the mother and her son, they observed the baby trying repeatedly to nurse and Joansie doing her best to

147

help. However, when the infant's hand, which had been tightly grasping his mother's chest, seemed to go limp, it was a positive sign that he was weak and needed to be nourished in the zoo nursery. When Roscoe joined Kahn in the nursery, he weighed just two pounds fourteen ounces.

Although both animals began eating well and appeared healthy after their first few days in the nursery, Kahn came down with a bad case of diarrhea and suddenly stopped taking his formula from a bottle. His body temperature dropped dramatically. He responded to intravenous feeding, but he suffered a relapse when he was put back on the bottle. Dr. Gardner suspected he had contracted a virus. As the vet staff prepared antibiotics, Kahn suddenly stopped breathing. Using artificial respiration, Gardner was able to revive him. However, Kahn was obviously very sick, and Hanna and Gardner decided his condition was serious enough to warrant a trip to the Columbus Children's Hospital's emergency room.

Columbus Children's Hospital has an active but low-profile animal research program. Primates are among the animals used in research aimed at improving human health care. Hanna was aware of Children's animal research program because of his daughter Julie's fight with leukemia. Despite his obvious love for animals and all forms of animal conservation, Hanna has stated publicly that he is grateful for animal research because he knows his daughter is among the human beneficiaries of such scientific study.

At the time of Kahn's illness, the hospital's animal research committee chair was Richard McClead, a pediatrician with an academic practice emphasizing research and education. When the doctor heard that a gorilla from the Columbus

Zoo was on its way to the emergency room, he went to see if he could assist with the animal's care. He waited for nearly an hour in the emergency room, but the gorilla never arrived. Still curious and hoping to lend his assistance, he returned to his office and called the zoo to find out what had happened. He was put through to Hanna, whom he had never met. Hanna explained that Kahn had stopped breathing again on the way to Children's, so he and Gardner had decided to stop at Riverside Hospital's emergency room, which was closer to the zoo. Sadly, the animal had died before they could get him to Riverside. Gardner believed that the gorilla had died from the effects of a bacterial infection, which was later diagnosed as salmonella. McClead expressed his condolences and told Hanna to feel free to call him if he ever needed any help with other infant gorillas. Hanna thanked the doctor for his offer and assured him that the zoo's other infant was doing fine.

The next day McClead received a call from the zoo: the second newborn had suddenly come down with a bad case of diarrhea. Fearing that Roscoe was suffering from the same bacterial infection that had killed Kahn, Hanna asked the doctor if he would come to the zoo and take a look at him. Although Roscoe did not appear to be in grave danger, McClead thought he needed to be rehydrated with intravenous fluids until he could build up his strength. The zoo had the equipment to provide intravenous feeding, but McClead was concerned about the animal's extremely low weight, probably a consequence of being born prematurely, and thought it best to admit Roscoe into Children's Hospital, where he was taken to the basement of Ross Hall, the hospital's animal research facility. (The name Roscoe was a combination of *Ross* Hall and *Col*umbus.) McClead and the zoo staff expected the infant male's stay in the hospital to be very short. Although there was no available literature for McClead to rely upon when it came to infant gorilla care, he was fairly confident that the similarities between humans and gorillas would allow him to the treat the animal the way he would any of his other patients. He would later say that the only times he got in trouble during his treatment of Roscoe were when he deviated from caring for him as if he were human.

McClead discovered that Roscoe was suffering from a primate-specific strain of salmonella, probably the same bacteria that had killed his half-brother. Meanwhile, back at the zoo, sixteen-month-old Cora, who was also living in the nursery, had stopped eating. The zoo feared there had been some sort of bacterial outbreak, potentially carried by, but not contracted by, Toni or Joansie.

As McClead proceeded with Roscoe's care, he soon found out that inserting an intravenous tube into a gorilla is a lot different than putting one into a human infant. "With human babies, you just have to worry about the arms. Roscoe knew how to use his feet. He really gave us a battle," McClead says. Volunteers from both the hospital and the zoo stepped forward to help with the gorilla's care. A team was set up to monitor the infant's progress at all times. Within two weeks, Roscoe was doing well enough to take him off his IV and feed him an ounce of formula containing sugars, protein, and other nutrients every three hours. In anticipation of his return to the zoo, the staff from the zoo nursery and the hospital set up a special treatment facility for Roscoe adjacent to the nursery. Cora's condition had improved, and it appeared the bacteria scare was behind them. The keepers decided that Cora's loss of appetite might have been caused by a sudden lack of at-

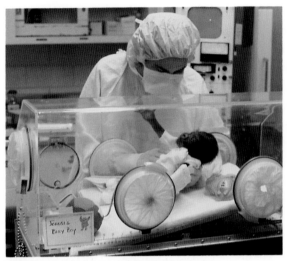

Roscoe at Columbus Children's Hospital

tention from her keepers. She had been used to having the nursery staff's undivided attention, but with the sudden addition of two other gorillas to the nursery, and the subsequent failing health of the two newborns, her playtime had been cut drastically.

Roscoe returned to the zoo on September 1. Visitors to the zoo, and the nursery staff, were thrilled to have him back. Several get-well cards adorned his nursery area, sent to him by concerned Columbus residents. On his first day back, Roscoe suffered a small and seemingly insignificant cut on his hand from a thermometer inside his incubator. The keepers wouldn't have given the cut a second thought, except that it wouldn't stop bleeding. The cut continued to bleed slowly for several days. At the same time, Roscoe's health seemed to be declining. When the vet staff administered blood tests to monitor his condition, the bleeding problem occurred again. They couldn't get the area where the needle was inserted to stop bleeding. In addition, Roscoe's temperature and heart rate dropped dramatically, and he began vomiting. Hanna called the hospital and put the gorilla in his own car to transport him back to the emergency room. On their way to the hospital, Roscoe's condition worsened, and Hanna decided he needed immediate help. He stopped at a local firehouse, where paramedics put Roscoe on oxygen and transported the gorilla to the hospital by ambulance. In addition to the bacterial infection, Roscoe was found to be suffering from anemia. Dr. McClead knew that anemia was a sign of a chronic disease, but he hoped it was merely a side effect of the medication the gorilla had been given for the salmonella. McClead treated the anemia by giving Roscoe additional iron. Unfortunately, Roscoe refused to take the iron through his mouth. The doctor was forced to administer it through a painful injection into the gorilla's deep muscles. As predicted, the iron helped build up the animal's red blood cell count, which aided in clotting, but it affected his level of antioxidants, causing him to suffer a vitamin E deficiency. The doctor could see the gorilla's ailments escalating, but he remained confident that Roscoe could be successfully treated and returned to the zoo.

After a week at the hospital, Roscoe returned home again. His weight had gone up to nearly four pounds. On his arrival, the zoo's head nursery keeper, Dusty Lombardi, told a local newspaper, "He looks 100 percent better." The hospital and zoo had worked out a twenty-four-hour care regimen that included the gradual introduction of milk and some other foods. But the happiness and optimism generated by the gorilla's return to the zoo were short-lived. The introduction

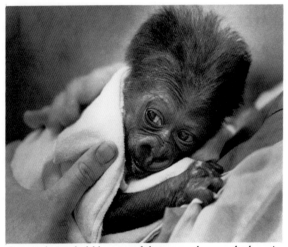

Roscoe being held by one of the researchers at the hospital. Credit: United Press International.

of food caused Roscoe to again suffer from severe diarrhea. He had to return to the hospital. McClead decided he needed to find out if the bacteria had damaged the gorilla's intestines. He asked one of the hospital's gastroenterologists to take a culture of the gorilla's intestinal lining. If the intestine was normal, the doctors would see long, fingerlike cilia fibers, which help break down and digest foods, as they looked at the culture through the microscope. Instead, they saw very few cilia fibers. McClead says Roscoe's culture resembled those taken from children in Third World countries who have suffered from chronic diarrhea. There were virtually no digestive enzymes present to break down food.

The doctor decided to give Roscoe's bowels a rest and put him back on intravenous feeding. After several weeks, the doctors took another biopsy of Roscoe's intestine. This time, they found that it looked almost completely normal. The treatment appeared to have been a complete success. It was expected that Roscoe would begin to eat regular food and that in short order he would return to the zoo. However, yet another complication surfaced. Roscoe refused to take any food by mouth. McClead had seen similar symptoms in human infants. He says that doctors have a certain time period in which they must introduce newborns to feeding through a bottle. When that passes, it is difficult to get them to eat normally. Since Roscoe kept pulling out his IVs, McClead had doctors insert a permanent feeding line into his stomach. Although the feedings worked and Roscoe started to gain weight, they were very expensive. The tube-feeding formula cost the hospital about $1,500 per month. Since it appeared that Roscoe would not be returning to the zoo for a while, the zoo and the hospital had to reach an agreement about covering the hospital's expenses for his care. McClead estimated that it would take six to eight months of constant hospital care if Roscoe were to have a chance to recover.

What had at first appeared to be a short-term and simple task for McClead and his team had turned into a long-term commitment, but the doctor never considered bowing out. Instead, he persevered and continued to rely on the dedication of the volunteers. "I consider the zoo a real community resource," McClead says. "I knew I was caring for an important member of the zoo family, therefore I was a trustee of that community resource. I was not about to make any irresponsible decisions. I tried to help the zoo think through all of its options and was committed to caring for Roscoe as well as we possibly could."

McClead's commitment was apparent during the Christmas holidays. For the first time since Roscoe's arrival, the doctor had a difficult time finding volunteers to stay with him. Given no other choice, he decided to take the animal home himself. Although Roscoe required nearly constant monitoring, his presence turned out to be a highlight of the McClead family's Christmas. His three children, ranging in age from four to seven, couldn't get enough of their holiday house guest. "All these years later, my kids still talk about that Christmas," McClead says. "I think it's the most memorable Christmas my family has ever had." Although his kids were thrilled to have a gorilla in the house, McClead's wife didn't share their enthusiasm. "She told me, 'You're going to have to take care of him! I can't be his mother, too.'" When the word got out among McClead's family that he had an unusual Christmas visitor, the McClead home was suddenly flooded with other unexpected house guests. "I saw relatives I didn't even know I had," he says.

As the weeks and months passed, Roscoe became a well-known patient at the hospital. The hospital staff was enjoying the novelty of caring for their wild patient. Many of them walked the hospital's hallways proudly wearing t-shirts that read "Gorilla Watch." Although the hospital continued to donate all of its staff time, such services as lab tests, supplies, and respiratory equipment, in addition to the high cost of the tube feeding, were threatening to bankrupt the zoo's medical budget. As the bills for Roscoe's care mounted at a rate of about $2,000 a month, Hanna knew that his medical budget—just $18,000 annually to care for the zoo's nearly 8,000 specimens—would soon be depleted and that he would have to locate additional support. Early in 1981 Hanna decided to ask the public for donations. It was not a decision that came easily. He worried that some people might be offended by the request. Hanna also told reporters that he asked himself, "Why am I helping raise money for animals when my own daughter has leukemia? Why am I not out there raising money to save kids?" Although Hanna's daughter's leukemia was in remission, she still had to receive bone marrow tests every six weeks. He said it was his daughter who gave him the peace of mind to ask the public for money to save Roscoe. "When I get home every night, the first thing Julie asks is 'How's the baby gorilla?'" he said. (In fact, Hanna had raised money for leukemia research by leading a campaign that netted $16,000 for the National Leukemia Foundation and had served on the board of directors of the foundation's Central Ohio chapter.) "We have a chance to save Roscoe," he stated. "I don't want to make a practice of [asking for money], but this money is going to save his life."

Hanna held a press conference and asked for $26,000 over a period of eight months. The money would pay for the care the hospital had already provided and for the future care that appeared necessary. He explained that anyone donating to the Roscoe Gorilla Fund would receive an autographed photograph of the seven-pound, six-month-old gorilla. In the photo, Roscoe was wearing a shirt that said: "I'm a Good Kid." When asked what he would do if the money couldn't be raised, Hanna replied that something at the zoo would suffer. "Gorillas and all great apes have been used in research, and given their lives toward the cures for cancer and all kinds of diseases," he said. "It seems like this is one way to repay the animals for everything they've given us."

Immediately, the zoo's phones began ringing as people pledged their support for Roscoe. A local elementary school

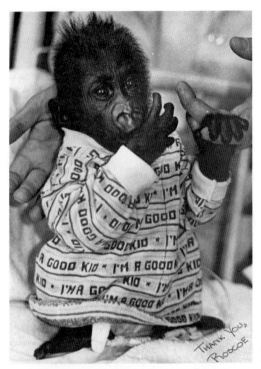

The "autographed" photo of Roscoe given to
contributors to the Roscoe Gorilla Fund

made plans to sell popcorn fritters for two months, with all
the money going to the Roscoe fund. A twelve-year-old boy
pledged $30, saying he would do odd jobs at a local grocery
store during the weekends to raise the money. About $1,200
was collected on the first day of the fund drive. In just a
month, nearly $20,000 had been collected.

As Roscoe's gorilla fund was growing, so was Roscoe.
By March he weighed more than twenty pounds. For Dr.

McClead, Roscoe's growth was complicated by the fact that he was getting a mouthful of teeth. Although Roscoe had shown great affection toward his nurses and caregivers, McClead often had to be the bad guy—the one who administered painful treatments. The doctor says that when he entered Roscoe's room, Roscoe would often bare his teeth and show his anger. Some of McClead's colleagues received minor bites from the gorilla, but luckily nobody was seriously bitten. However, in the back of his mind McClead knew that as Roscoe approached the weight of twenty-five pounds, the problem would only get worse.

By this time Roscoe had suffered through multiple bouts of infection around his feeding tube. The tube was inserted under his clavicle, and bacteria were continually invading the area. When he got an infection, doctors would have to stop the tube feeding and give him antibiotics. To McClead, it seemed that there was nearly an endless cycle of infection, treatment, and restarting of the feeding process. In the spring McClead received a call from the nutritionist who was handling Roscoe's feedings. She said he was lethargic and didn't look right. His hair was lighter and he had several lesions on his body. McClead diagnosed the condition as a zinc deficiency and tripled Roscoe's zinc dosage. Within a day, the nurses could see an improvement. Within a week, all the lesions were gone. It was another small victory, but Roscoe was still not progressing well.

McClead knew he couldn't keep Roscoe on total parenteral nutrition for much longer. The gorilla was already appearing jaundiced, a sure sign of liver damage. As his liver deteriorated, McClead knew the end was near. He had already escalated the treatment far beyond what was originally intended, and treatment of the liver problems would probably lead to other complications. After weighing all the options, McClead reluctantly decided not to treat Roscoe further, and his health deteriorated rapidly. On July 29, 1981, just two days before his first birthday, Roscoe died in the arms of Columbus nursery keeper Criss Pendleton. Of all the nurses who had cared for him over the previous months, Criss had spent the most time at his side. Pendleton reportedly just sat with him and cried. "He had such a personality," she said at the time of his death. "I will always remember him as a friend." Dr. McClead asked Dr. Gardner, the zoo veterinarian, to perform Roscoe's autopsy. The autopsy showed that all of Roscoe's organs were severely jaundiced; it was clear that the medical team had done all that it could and that McClead had made the right decision.

The zoo had collected $23,000 for the Roscoe Gorilla Fund, almost exactly enough to break even. But for both the hospital and the zoo, this had been much more than a break-even experience. Roscoe's struggle for survival had led to a strong relationship between the two institutions. McClead, for example, has since been involved in every gorilla birth at the zoo. He and his staff have helped train the zoo staff in the care of infant gorillas. Also, McClead and a Children's Hospital epidemiologist walked around the zoo with the keepers and veterinary staff and made suggestions for keeping conditions as sanitary as possible when it is necessary to hand-rear an animal. Further, as a result of the experience with Roscoe, a predelivery protocol has been established, which involves a preparation and strategy session with doctors, keepers, and staff. The group goes over all the equipment in the hospital and nursery and determines how far they will prolong medical care if another situation like Roscoe's arises.

Lessons learned from the experience with Roscoe were put

into practice two months after his death when Zura was born to Toni and Oscar. At the age of six weeks, the gorilla began having difficulty breathing. Her lungs had filled with blood from a heart defect. With the help of Dr. McClead and a heart specialist from Children's Hospital, the zoo staff was able to treat Zura successfully. "Because of Roscoe, those people were able to help Zura, who is now living a healthy life at another zoo," McClead says.

Just two days after Roscoe's death, on what would have been his first birthday, Roscoe's mother, Joansie, gave birth to her second male offspring, O.J., weighing four and a half pounds. Although the infant was eventually sent to Buffalo as part of a breeding loan agreement, his birth proved that the Columbus gorilla program had persevered.

8

Jumoke: *Gorilla, gorilla, gorilla*

Although the Columbus Zoo has had a remarkable record of gorilla births, the keepers are quick to point out that pregnancies and births are not the focus of their program. Rather, they are the positive results of a program that strives to replicate the social setting and interactions experienced by gorillas in the wild. If gorilla births were the primary goal, every newborn would be pulled immediately from its mother, because female gorillas are capable of reproducing every year if they are not responsible for raising their offspring. In the wild, the rate of reproduction for healthy females is every three to five years. Although some Columbus females, such as Toni, have given birth with great frequency, this has not necessarily been the staff's preference. In fact, every effort has been made to have Toni keep and raise her offspring, but unfortunately she has never been able to do so.

Toni's sixth baby, Jumoke (Swahili for "loved by all"),

was born November 10, 1989, to her and Sunshine. Three
gorillas had been successfully mother-reared in Columbus by
the time Jumoke was born, and the keepers hoped that their
experience with the other animals would allow them to help
Toni finally succeed. During Jumoke's first three days of life,
the keepers confirmed several successful feedings, and the
newborn appeared to have good skin color and healthy, clear
eyes. The keepers cautiously let things progress, keeping in
mind that Toni had a history of seeming to grow uninterested
in or frustrated by the process of raising her newborns.
Norman, a male born to Toni just seventeen months earlier,
had died suddenly and inexplicably at the age of ten days
while in Toni's care. The keepers knew that their desire to let
Toni rear her own offspring would have to be tempered by the
knowledge that she literally had a life in her hands.

On Jumoke's sixth day, the veterinary and keeper staff
noted that the newborn seemed aware of her surroundings
and had a strong grip; her eyes remained clear. However, as
the day progressed, Toni began acting as she had with previous infants: instead of holding Jumoke and making eye contact, Toni wasn't looking at her very much, and she was
handling the newborn in a seemingly careless fashion. She
was also making no effort to have the infant breast feed, and
twelve hours passed between confirmed feedings. Dr. Mc-
Clead was summoned from Children's Hospital to help assess
the situation, and he and the veterinary staff determined that
Jumoke was not getting enough nourishment. The decision
was made to intervene and take her to the nursery. This time,
however, the keepers planned to keep the infant in the nursery
for just forty-eight hours. After normalizing her fluid and
nourishment levels, they planned to reintroduce her to Toni,

Jumoke as an infant

hoping that she would again accept her infant. Perhaps all she needed was a two-day break from the newborn.

The keepers were able to coax Toni away from her infant without anesthetizing her. The sight of the veterinarian's equipment was enough, and she put the infant in a clump of wood-wool and walked into a chute leading to an adjacent cage. When the keepers rushed Jumoke to the nursery, they found that her temperature had dropped to ninety-two degrees. This was a cause for concern, because, McClead notes, gorillas protect themselves from the effects of malnourishment or sickness by reducing their body temperature.

As Jumoke's temperature slowly returned to normal and her vital signs stabilized, the gorilla staff began preparing Toni for the reintroduction of her newborn. Since the keepers would have to feed Jumoke even after she was returned to her mother, they needed a way to gain regular access to her. They devised a strategy involving two bottles—one that would be fed to Toni when she brought Jumoke to the mesh, and a second that would be given to the infant. As Jumoke grew stronger in the hospital, Toni was fed from a bottle, while keeper Charlene Jendry held the second bottle lower to the ground, as if she were feeding the baby.

Within twenty-four hours of being pulled and taken to the nursery, Jumoke was brought back to the ape house in an attempt to reassure Toni that her offspring was doing fine. Within forty-eight hours, the keepers replaced Jumoke in a pile of wood-wool near the spot where Toni had left her and, hoping for the best, opened the door to the chute giving Toni access to the cage. Toni's immediate reaction to Jumoke would determine the newborn's fate. As Toni entered the cage, she walked over to Jumoke. She leaned down and smelled the infant, then turned her back and walked to the edge of the cage. The reintroduction had failed.

Jumoke was returned to the nursery, where a new set of gorilla-rearing protocols was in place. First, the infant would not be put on public display. Although baby gorillas have always drawn large crowds to the zoo, putting the newborn on display would only detract from the task at hand—raising Jumoke in the most natural way possible. The long-term goal was to introduce her to a surrogate provider within eight months.

At the age of four and a half months, Jumoke was riding on the back of nursery keeper Barb Jones, who was implementing the lessons she had learned from watching Columbus's experienced gorilla mothers. To supplement the gorilla vocalizations being played over the nursery sound system, Jones imitated other vocalizations during feedings, playtime, and rest periods. Jumoke was also taken to the gorilla house on a regular basis, where members of Mumbah's group were allowed to come to the mesh and interact with the infant. When Jumoke was approached calmly, Jones would allow the animals to touch each another. Jones was especially intent on observing the behavior of the females, hoping to identify an appropriate surrogate for Jumoke.

As Jumoke approached her eighth month in the nursery, the keepers were optimistic that they would soon be able to introduce her to a surrogate in Mumbah's group. However, just before the introduction was to take place, Bongo died. Bongo's four-year-old son, Fossey, and the two females who were part of his group, Molly and Sylvia, were now without a silverback, and they would have to be introduced to Mumbah's group first. The unexpected change in group dynamics delayed Jumoke's introduction by eight months.

By the time the keepers thought the group was ready for Jumoke, she was almost a year and a half old and had grown attached to Jones and the daily attention she received in the nursery. The scientific name for the family, genus, and species of the western lowland gorilla is *Gorilla, gorilla, gorilla.* Despite the best efforts of the gorilla and nursery staff to provide her with care that was as true to the wild as possible, Jones likes to say that after sixteen months in the nursery, Jumoke was more like a *Human, human, gorilla.* Her attachment to the life she had led with humans promised to make her introduction to a surrogate particularly difficult.

Sylvia, one of the females in Mumbah's group, had shown the most interest in Jumoke. In most cases when an infant is being introduced to an adult, keepers watch for acts of aggression that might harm the younger gorilla. In Jumoke's case, however, the adult seemed to be at risk. While Jumoke aggressively attacked and vocalized angrily, Sylvia showed great restraint. After a few hours, things had calmed down enough to call the introduction a success.

Sylvia had come to Columbus in 1986 from the Baltimore Zoo via the National Zoo in Washington, D.C. While the keepers at both of her previous zoos spoke highly of her warm personality, at age twenty-two she was described as human oriented and lacking the skills a gorilla needs to succeed in a social group. She seemed destined to live her life in captivity as a subordinate female. "Sylvia came to this zoo without many gorilla skills," Jendry says. "All her home zoo asked was that she someday be able to socialize with just one other gorilla. Obviously, she's done more than that, but it required time and space and not rushing the situation and letting Sylvia figure out how she fit in." Sylvia had originally been placed with Oscar and Pongi. Because the animals lived in Oscar's Yard, removed from the base operations of the keeper staff, Sylvia was no longer in constant contact with humans. She was soon interacting with the other gorillas and within a year was observed breeding with Oscar. She was able to watch Pongi giving birth to Colbridge and to participate in the rearing of the infant.

In 1990 Sylvia was taken to the main great apes building and introduced to Bongo, who was living with Fossey and another female, fourteen-year-old Molly. In the six months Sylvia lived with Bongo before his death, the pair was never observed breeding. However, Sylvia held her own in skirmishes with Molly, and after Bongo's death Fossey seemed to show a closer attachment to Sylvia than to Molly. Fossey and Sylvia were introduced together to Mumbah's group, where Fossey's presence with Sylvia enhanced her social standing. As a provider, Sylvia became the group's dominant female, holding her own during brief aggressive exchanges with other females. Maintaining her dominant role among the females, Sylvia defended the juveniles against the aggression of the other animals, and other females began coming to her defense when fights occurred. To the surprise and delight of the Columbus keepers, Sylvia had come full circle, from a humanized and subordinate female to a highly social and dominant member of a large group.

Sylvia's status and apparent confidence enabled her to act successfully as Jumoke's surrogate. With Sylvia's help, Jumoke gradually became an active member of the group. Although Jumoke remained attached to Jones and other human keepers, Jones says it wasn't long before she considered Jumoke a *Human, gorilla, gorilla,* well on her way to becoming a full-fledged *Gorilla, gorilla, gorilla.*

Annaka, an impressively large eight-year-old blackback, arrived in Columbus on breeding loan from the Philadelphia Zoo in October of 1993. He was introduced to Mumbah's group, joining Sylvia, Jumoke, and nine other gorillas. As the offspring of the Philadelphia Zoo's famous breeding pair, John and Snickers, Annaka represented a new and important bloodline in Columbus. His health and well-being became more important than ever two years later, when a tragic fire at the Philadelphia Zoo killed the entire gorilla collection and three orangutans. Among the six gorillas that died of smoke inhalation were John and Snickers and three of Annaka's siblings.

Before her sixth birthday, Jumoke was observed breeding with both Mumbah and Annaka. "Annaka was a young interloper in Mumbah's group," recalls head gorilla keeper Susan White. She observed growing tension between silverback

Annaka peering through the mesh of the gorilla villa

Mumbah and the brazen blackback as the months passed and sexual relations between Annaka and Jumoke continued. To avoid a potentially violent encounter between the two males, Jumoke and Annaka were separated from Mumbah's group in April of 1996. After just one month alone with Annaka, a test showed that Jumoke was pregnant.

That same spring, construction was begun on a new $5.7 million gorilla habitat that would expand upon the gorilla villa and ape house, but the construction created short-term space constraints. Sunshine's group, which at the time included Toni and Colo, was moved temporarily to the Detroit Zoo, which had been without gorillas for nearly fifteen years. Colo's status in Columbus and her age (approaching forty) led keepers to decide that she should not be moved but would join Jumoke and Annaka instead. They thought the arrangement would be best not only for Colo but also for Jumoke and Annaka. Colo's presence might help the animals mature.

Immediately after her introduction to Annaka, Colo set some boundaries. Reacting to what the keepers agree was inappropriately aggressive behavior, Colo threw feces at Annaka, hitting him squarely in the face. The keepers say this was a rare and bold gesture, and Colo apparently made her point. She and granddaughter Jumoke seemed comfortable together, often protecting each other against Annaka's aggression. Annaka's behavior had, in fact, concerned the great ape staff from the beginning of Jumoke's pregnancy. They had hoped that Annaka and Jumoke could be together twenty-four hours a day as the pregnancy progressed, and that Annaka could be present at the birth. By May, according to White, Annaka, now eleven years old, seemed to have proved to the keepers that he was ready to stay with Jumoke and Colo overnight. However, when they arrived at the ape house the following morning, they found bite marks on Jumoke's hands and feet. For the remainder of the pregnancy, Annaka was only allowed to share a cage with Jumoke and Colo during the day when his actions could be closely monitored by the staff. "He has it in him that he must manhandle and dominate submissive females, which is either immaturity or simply inappropriate," White says. Charlene Jendry and Beth Armstrong recall that both Oscar and Sunshine, two other males who sired offspring at a young age, went through similar stages of "immaturity." According to Armstrong, it wasn't until Sunshine reached the age of fifteen that "he seemed to figure it out."

Meanwhile, the keepers worked with Jumoke to help prepare her for the birth of her first offspring. Considered a highly intelligent gorilla by her keepers, Jumoke was put through the most extensive "motherhood training" the zoo

had ever attempted. Since Jumoke had never witnessed a gorilla birth or the interaction between a gorilla infant and its
mother, Barb Jones and keeper Debbie Elders were determined to teach her everything they had learned from observing other successful gorilla mothers. The program started,
Elders says, with the confirmation of trust between the gorilla
and the trainers; without it, very little can be accomplished.

Some of the training was simple—for example, Jones used
a stuffed gorilla toy to demonstrate the proper way to cradle
and gently handle a newborn. The keepers also introduced
Jumoke to basic vocal commands, hoping to be able to
"coach" her when labor began. Other aspects of the training
were more difficult and time consuming. For example,
Jumoke was gradually introduced to the probe that the veterinary staff would use to perform an ultrasound. The keepers
hoped Jumoke would grow comfortable enough with the
probe to allow the staff to perform an ultrasound without
having to anesthetize her. They began wearing the ultrasound
wand around their necks during feedings, allowing Jumoke to
touch it, smell it, and even taste it.

Although the gorilla was responsive to the training, the
keepers eventually ran out of time, and she had to be anesthetized for the procedure, which was performed in October. As
Annaka stood atop a tractor tire in his cage craning his neck
to get a clear view, Jumoke was returned from the hospital to
the ape house to recuperate. As Jumoke slowly came out of
the anesthesia, Jones reached through the mesh, softly stroking her and trying to help her remain calm. Jumoke was shiv-
ering from the effects of the anesthesia, so Jones and head
keeper Susan White placed an electric heater next to her cage
to provide some comfort. White also did her best to comfort
Colo, who seemed uneasy with the separation from Jumoke.

White walked to the edge of Colo's cage carrying a video camera, intending to show her video footage of one of Jumoke's training sessions. As White rewound the tape for playback, she explained to Colo what was on the tape and asked, "Do you want to see Jumoke?" She stuck the camera's small eyepiece through the mesh for Colo to look through, and Colo reacted by sticking out her long tongue and tasting it. "Don't eat it! Look into it," White said. "It's Jumoke." Colo leaned over, her large head twice the size of White's, and gave her attention to the eyepiece. For a minute or more she stood still, watching the tape. Whether or not she knew that the small black-and-white image was Jumoke, the action on the tape kept her interest.

Based on the ultrasound, the veterinary staff determined that Jumoke would give birth around New Year's Day 1997. White gathered a group of volunteers who would take part in a round-the-clock birth watch program, scheduled to begin on Sunday, December 14. Such birth watches had been used in Columbus since Bridgette was pregnant with the twins, although the last two gorilla infants born at the zoo had arrived before the watches had even started. For Jumoke's first offspring, White didn't want to take any chances, and she got an early start. During the volunteers' orientation, she explained some of the visual signs of gorilla labor and went over some specific things the volunteers should look for. Since Jumoke was Toni's offspring, White also described some of the unique mannerisms Toni had exhibited during labor—such as a violent and seemingly uncontrollable shaking of her hands and arms—that might have been passed on to Jumoke.

Teams of two monitored Jumoke from an education classroom in the arthropod building next to the ape house. Using two remote-controlled cameras, the volunteers took turns working four-hour shifts, beginning at 4:00 each afternoon and ending at 7:00 in the morning with the arrival of the keepers. Night after cold winter's night, the volunteers weren't able to witness much more than Colo and Jumoke resting comfortably. Desperate to report any kind of news, they wrote things like "saw a mouse" and "are you sure she's pregnant?" on the log sheets. Colo's fortieth birthday was celebrated on December 22, followed by Christmas and New Year's Day, but Jumoke was showing few signs that she was about to give birth. The zoo extended the birth-watch schedule into January. There were a couple of false alarms when Jumoke, in a heightened state of excitement, paced the floor of her cage as if labor were beginning, but it was not until the night of January 26 that White became confident that birth was imminent. Jumoke slept comfortably through the night, but when White arrived the next morning, she found her in the early stages of labor. By 10:00 in the morning, Jumoke was in strong labor, and the veterinary, nursery, and gorilla staffs were prepared for Columbus's first gorilla birth in three and a half years.

At 1:12 that afternoon, Jumoke gave birth to a male infant. From the beginning, she was an attentive and protective mother. "Jumoke is handling the baby like an experienced mother," wrote Living Collections director Dusty Lombardi in the zoo's newsletter. "She's being incredibly gentle and caring. She knows exactly how to the hold the baby, she's making eye contact. She's doing more than any of us ever expected."

Jumoke allowed the keeper and veterinary staff to keep a close eye on the infant when he suffered a case of thrush and a mild respiratory infection. He did not have to be taken to the

Infant Jontu being held in Jumoke's arms

nursery for care. Jumoke allowed the veterinary staff to take the infant's temperature, listen to his lungs, and administer medicine. Jones and Elders showed great pride in her success as a mother. "I'd like to give the training all the credit for the success we've had with Jumoke and the baby, but the truth is no amount of training is going to convince a mother to let us touch her baby," Elders says. She believes that since Jumoke had grown accustomed to allowing the keepers and veterinary staff to touch her body during her pregnancy, she probably felt comfortable with those same people touching the baby. "This has come out exactly the way we all hoped," says Beth Armstrong. "How do we prove that it's the result of our preparation—the socialization and the husbandry process? I'm not sure we can, but it's exactly what we hoped for." The keepers decided to name the male Jontu, in honor of his deceased grandfather, John.

Meanwhile, Annaka has continued to disappoint the keepers with his behavior—"inappropriate at least, dangerous at

Jumoke

worst," White calls it. Determined to give him a chance to
take part in the rearing of his offspring, the keepers allow
closely watched visits between Annaka, Jumoke, and Jontu.
But White says Annaka's overly aggressive behavior toward
Jumoke continues. "He'll make nice hoots and vocalizations
before and after he goes in" with Jumoke, White says, "but
when they're together, it all happens again. It's not only confusing to us as we try to make decisions, it's also confusing to
the animals," who haven't been able to establish a routine because of Annaka's behavior. "Having an infant should have
been a very humbling experience for him," Armstrong says.
"It has been with other males we've observed. The reality is
Jumoke, as the mother of a newborn, should be able to call
the shots with any animal in the group, including the
silverback—and Annaka's not even a silverback. He's not
making the connection that there is something at stake. It is in
his best interest to adhere to what she wants and take care of
the baby. It's a very frightened vocalization that she makes.

He does not understand that he should not step over the line." As a result, Annaka's visits with Jumoke and Jontu have been short, sometimes less than thirty minutes, depending on how he behaves. "If Mumbah acted the way Annaka does, that surrogate group would not exist," Armstrong says.

The contrasting behavior of the young mother and father—Jumoke seems to be doing nearly everything right, while Annaka seems to be doing nearly everything wrong—illustrates the keepers' point that the gorillas are individuals, and this understanding dictates how decisions are made in the ape house. Jontu is a living example of the fact that there is more than one way to approach pregnancies, births, and social interactions. Once again it is clear that the key to moving the gorilla program forward lies in interpreting the animals' behavior in a flexible and intelligent manner.

9

Risks, Rewards, and the Future

Even the most casual observer can see how much the gorilla breeding, birthing, and socialization processes have evolved in the forty years between the births of Colo and Jontu. Breeding Mac and Millie was based on little more than a hunch; its success had as much to do with luck as it did with knowledge. Colo was born into an atmosphere dominated by uncertainty, where decisions were based on fear, excitement, and raw veterinary instinct. Jontu, however, was born to a well-prepared and socialized mother. His environment was a controlled and sophisticated one, thanks to the pioneering work of field scientists like George Schaller and Dian Fossey and the lessons learned from the combined successes and failures of breeding programs in Columbus and elsewhere. The sciences of gorilla breeding and socialization have, in fact, become fairly well understood, but only because individuals and institutions have been willing to go beyond conventional thinking in the interests of conservation, education, and, in some cases, simple curiosity.

From the beginning, risk takers have been the among the most influential architects of the Columbus gorilla program. Earle Davis purchased Columbus's original gorillas sight unseen, and brought them to Columbus hidden in orange crates aboard a passenger train. Warren Thomas permitted Mac and Millie to breed and made the first captive gorilla birth possible. James Savoy chose to treat the Columbus gorillas when they were diagnosed with tuberculosis rather than destroy them. Veterinarians like Harrison Gardner were willing to sedate and care for the gorillas when other zoos were afraid to do so. Dianna Frisch fought to allow brother and sister Oscar and Toni to breed. Doctors outside the veterinary field, like Nick Baird and Richard McClead, applied human medical techniques to the treatment of gorillas and prolonged or saved lives. Beth Armstrong, Charlene Jendry, and other keepers used unconventional methods to make the zoo's unique husbandry program a success. And, perhaps most notably, Jack Hanna fought to have the gorillas' quarters modernized, allowing them to experience the outdoors.

Hanna supported his gorilla staff and enabled them to implement creative methods to improve the gorillas' lives. Columbus Zoo Board president Dr. Nick Baird says it was Hanna's combination of an innovative spirit and a willingness to empower his staff that allowed the gorilla program to evolve into one of the best in the world. "Micromanagement can be the death knell of a program," Baird says. "If you are unwilling to take chances you are destined to be mediocre. You've got to put the power in the hands of the people who know the animals. You also have to have enough staffing with enough time to devote to making it right. If it's punch in, clean cages, and punch out, your program is doomed to be nothing more than an exhibit program."

The atmosphere created by Hanna and extended under the direction of his successor, Gerald Borin, has allowed important ideas that were born out of informal discussions among keepers to be implemented. "Almost everything that has happened as far as new idea generation, from a keepers' perspective, has happened either in the ape house kitchen, or around the picnic table outside," Jendry says. She emphasizes that problem solving or discussing the future of specific gorillas or social groups is an ongoing process that requires constant rethinking and updating.

Another of Hanna's major contributions to the gorilla program was his acknowledgement of the seemingly renegade practices of Howlett's Zoo founder John Aspinall. Called a "kook" by many zoo directors as late as the mid-1980s for creating large, age-diversified, social gorilla groups, Aspinall was ahead of his time when it came to gorilla socialization. Conventional wisdom said Aspinall was putting his valuable animals at risk, exposing them to one another's aggression in a "survival of the fittest" scenario. Hanna saw it differently. He came away from firsthand observation of Aspinall's Howlett's Zoo Park in England challenged by what he saw, proclaiming, "We have to take more chances" in Columbus. The result was a renewed commitment to support mother-rearing of offspring and the construction of Columbus's "gorilla villa." This gorilla habitat was controversial in its own right. "When the outdoor habitat was being built, the prevailing thought at the time was for zoos to go to 'naturalistic' exhibits," says Jendry. The villa's mesh sides, hanging ropes, and metal stoops and sitting areas were in direct contrast to the naturalistic movement. "The Columbus Zoo chose to go with an exhibit that had proven successful for Aspinall insomuch as he had bred and exhibited large groups at Howlett's. We went in that direction to give our gorillas a complex environment and a three-dimensional exhibit, and no one has been sorry," she says.

The visit to Howlett's also inspired the formation of Columbus's age-diversified groups, the largest and most important of which has been led by Mumbah, a silverback received on loan from Aspinall. "In the mid 1980s the Columbus Zoo gorilla staff came to the conclusion that a commitment to the early socialization of all infants, including hand-reared, was in the best interest of the next generation of gorillas. By July of 1986, we had had fourteen births at our zoo but none of these infants had been successfully mother-reared, leaving a gaping hole in our gorilla program," Beth Armstrong wrote in a presentation about their age-diversity program. "Thus began the commitment of our staff to create an age-diversified group through the use of alternative means beginning with surrogates."

An emphasis on mother-rearing and socialization subsequently outweighed the program's focus on simple breeding. In August of 1986, Bridgette and Bongo successfully raised Fossey, ending Columbus's string of unsuccessful attempts to allow its mothers to raise their own offspring. "Our philosophy was derived from the recognition that blame for many situations of maternal neglect lay not on the gorillas themselves but with the collective zoological institutions' lack of understanding of gorilla husbandry and behavior," Armstrong wrote.

> Often times a first time mother's first exposure to an infant was with the birth of her own offspring. In the case of multiparous females, a history of pulling their infants immediately after birth had helped reinforce a negative connotation associated with giving birth. In addition, due to a perceived need to exercise caution with newborns, the silverbacks were often separated from the females, thereby creating a change in routine that we believe created a more stressful environment. Births were often accompanied by isolation, frequent human visitors, changes in routine, and lack of privacy. One begins to see that in our attempt to be cautious we effectively created a strange and negative environment for mother gorillas.

While recognizing and emphasizing the importance of mother-rearing, the gorilla staff understood that there would be times when it would be impossible, because of medical or other complications. Therefore, the keepers drew up a plan to use surrogates to help with the proper socialization of nursery-reared infants. In other words, just because an infant had to be pulled and cared for in a nursery didn't mean it was destined to be raised by humans. It was clear that adult gorillas were capable of playing a necessary and vital role in the early socialization of infants. The keepers believed that the socialization process was essential to the infant; beneficial to the surrogate mothers and fathers, as well as to any juveniles in the troop; and important to future generations of gorillas in captivity. "We believe that there is a sense of responsible action inherent in rearing a youngster and that the inclusion of

Beth Armstrong and Charlene Jendry with Bongo, mid-1980s

infants affects all group members," Armstrong wrote. "It is this very concept that was the driving influence in creating an age-diversified group."

The zoo's first experience with a surrogate involved two somewhat older animals, Mac II and Mosuba. At the age of three, the twins were the youngest gorillas Columbus had ever integrated into an established group. The keepers decided to use eight-year-old Cora as more of a guardian for the twins than a surrogate. The twins were introduced to Cora first, and then the three were introduced to the group together. After the successful introduction, the keepers noted that Colo also took on a protective and mothering role with the twins. The following year, Colo became the zoo's first true surrogate, caring for fourteen-month-old, nursery-reared J.J. She played the role of mother flawlessly. "The first time we saw Colo lift J.J. up and put him on her back, the feeling was incredible," Jendry says. Colo's contribution seemed somehow appropriate, since her birth had done so much for the zoo and its gorilla husbandry program.

The program evolved even further when Lulu, a pregnant female, was introduced into Mumbah's group, where she gave birth and reared her female offspring, Kebi Moyo. Lulu's keepers had noticed that she was uncomfortable in the presence of her mate, Sunshine. Fearing that her uneasiness with the silverback might affect her ability to raise her child, the keepers decided to take a chance and move her to Mumbah's group. Three years later, the keepers were presented with a similar challenge when Oscar died, leaving his pregnant mate Pongi without a silverback. Bolstered by their previous success, the keepers introduced Pongi and her five-year-old male offspring, Colbridge, into Mumbah's group. Again, the introduction was successful, and Pongi gave birth and raised her female infant, Casode, within the group.

By 1994 the surrogate program had gained a national reputation and helped to increase Columbus's influence within the zoological community. Other zoos began sending infants to Columbus to be introduced to surrogates in an age-diversified group. Four-month-old Nia arrived from the Oklahoma City Zoo and, after four months of nursery care, was introduced to Sylvia, an experienced surrogate. In late 1996 Akanyi arrived in Columbus from Chicago's Brookfield Zoo. The three-month-old male was still recovering from several medical problems, including septicemia, a form of blood poisoning that had caused him to fall into a coma shortly after his birth. A combination of Columbus's reputation for gorilla medical care and its surrogacy program led the leaders of the Brookfield Zoo, which also has a well-respected gorilla program, to send Akanyi to Columbus. "Our name is out there," says head keeper Susan White.

Other zoos have also sent their keepers to Columbus for consultation and direction. Responding to many requests for the opportunity to train with the Columbus keepers, Columbus's gorilla staff created a free three-day training program that was implemented in the summer of 1997. Primate keepers from other zoos have the opportunity to observe the Columbus gorillas, meet with the gorilla and nursery staff, and review data related to the various gorilla programs. It is an unprecedented initiative, showing Columbus's dedication to the animals. "Zoos have traditionally been protective of proprietary information," Dr. Baird says. "It's neat to see the information sharing that is being led by the Columbus staff. Let's face it, we compete with the Cincinnati Zoo for visitors. But that doesn't mean the animals should suffer. If we're protecting information that could potentially help another zoo take better care of their animals for the sake of attracting visitors, then we don't have our priorities in line. This program is an example of how the animals are put before the bottom line."

When conferring with other zoos striving to build an age-diversified group, the Columbus keepers caution that there is some risk involved in executing multiple introductions of infants and pregnant females to unfamiliar silverbacks. They warn that the gorillas must control the decision-making process. If the animals don't exhibit the right behaviors, Columbus's methods should not be used. Proper cage configuration is also a vital element of successful introductions.

By providing escape routes and cages that do not have dead ends, the Columbus keepers say they have alleviated some of the tension inherent in the introduction process. "It is important to note that the use of surrogates and the introduction of already pregnant females has worked well for our institution but we would caution others to recognize that these accomplishments were risky and that when choosing this course the gorillas must exhibit consistent, appropriate behaviors in order to even be considered for alternative methods," Armstrong wrote in their husbandry brochure. Mumbah's patience and unique tolerance for the introduction of new animals into his group is credited for much of the success of the Columbus program.

When keepers from other zoos come to Columbus to see Mumbah's group, they observe the interaction of two infants (there will be three when Akanyi is introduced to the group), three juveniles, one blackback, and five adults. The keepers say that Mumbah's group is an excellent example of the kind of gorilla group that would be found in the wild. The presence of many gorillas of different ages provides stimulation for all the animals. Armstrong compares the group's dynamics to an active and healthy neighborhood. The adults are kept on the alert by the young ones, who are constantly stirring things up. There is the gorilla equivalent of laughter, play, confrontation, and discipline—all healthy elements of a functioning social group.

Meanwhile, beyond the borders of the zoo, Columbus has led American zoos in the formation of a group called Partners in Conservation, coordinated by Jendry. The group works with the Dian Fossey Gorilla Fund and the Mountain Gorilla Veterinary Project to protect gorillas in the wild. With the

Jendry with Rwandan guide Alfonse at the
Karisoke Research Center, 1992

professional and monetary support of the zoo, Jendry has
studied free-ranging gorillas at the Karisoke Research Center
in Rwanda. Partners in Conservation is working to help protect not only the extremely endangered mountain gorilla but
also the *people* who live in the region. By supporting an or-
phanage for children whose parents were killed in Rwanda's
bloody civil war, Partners in Conservation and the zoo recognize that the lives of the people and the animals are inter-
twined. It makes sense for the zoo to assist both.

"Zoos are evolving tremendously," Baird says. "The role
of zoos right now is to educate people about nature and the
wild. Existence is so fragile, many animals are on the edge.
Zoos are not Noah's Ark. It's unrealistic to think we can

populate the wild with captive-born animals. But zoos can provide the invaluable service of providing knowledge and appreciation." Gorillas are one of several endangered species that are flourishing in captivity. While gorilla populations continue to decline in the wild—despite heroic efforts to protect them, gorillas continue to be killed as crop pests in West Africa and illegally captured and killed by poachers—there is actually some discussion of an overpopulation problem in captivity. The Gorilla Species Survival Plan (SSP) of the American Zoological Association (AZA) states that capacity for captive gorillas in North America is about 400. At present there are about 350 gorillas in North American zoos, with a net growth of about 10 gorillas per year. "There is currently a population management issue, which centers on what do to with the abundance of young males [in captivity]," says Jendry. Since one adult male can lead a group containing several females, there is a greater need for females in captivity, but unfortunately there is a preponderance of males. "It would be nice if there were three females for every male, but that's not the way it is," says Armstrong. At least two American zoos, St. Louis and Cleveland, are attempting to address the problem of male gorilla overpopulation by forming all-male groups.

In its Gorilla SSP Master Plan, the AZA addresses the population issue by commending the enlightened trend toward the formation of larger social groups, which creates a lower ratio of exhibits to population size and promotes genetic diversity. Moreover, to ensure that all of the great apes are treated properly—not just the extremely popular gorilla—the SSP encourages zoos to "consider the construction or renovation of facilities that will serve other great ape species as well. Zoos that can accommodate only one great ape should choose orangutans or chimpanzees."

Armstrong would like the AZA to take the policy a step further and perhaps set standards to.help determine which zoos are capable of housing gorillas in appropriate habitats and which are not. She says that the gorillas' popularity can be detrimental to their care because even zoos that can't adequately accommodate the needs of the animals still want to display them. Zoos should have to prove they can provide such things as abundant bedding and a varied diet, and they should have programs in place that encourage the mother-rearing of infants and the formation of age-diversified social groups before they can acquire gorillas. Jendry agrees, saying it would be best for the species if the captive population were spread among the zoos that have the capacity and financial strength to house gorillas in large, age-diversified social groups, rather than keep them in small groups. Not only are large, age-diversified groups best for the animals, based on the knowledge gained from scientific study in the wild, they also address the issue of aged or nonbreeding animals. Older gorillas have been considered "surplus" by some zoos, but as Columbus has proved, they are extremely important in the creation of diverse groups and in the socialization of nursery-reared infants.

When discussing the future of the Columbus program—specifically the surrogate program—the keepers talk first about Mumbah. At age thirty-five, he is very mature for a captive male. "If Mumbah dies, that could be the end of the surrogate group," Armstrong says. Although they consider Macombo II to be his "heir apparent," based on Mac's proven skills around infants and his exposure to Mumbah's leadership and patience, the fact is there are few silverbacks with the temperament to accept the offspring of other silverbacks. There have been more cases of aggressive confrontation and even infanticide by silverbacks than of calm acceptance, such as Mumbah has shown toward other gorillas' infants for the past decade. "Even if it ended tomorrow, and the surrogate group disbanded out of necessity, at least we proved it could be done," Armstrong says. But with an optimistic view of the future she adds, "If this group continues on, even for three or four years, you could get a couple more infants." It's not out of the question for the zoo's two thirty-three-year-old females, Lulu and Pongi, to each have another offspring (both have given birth three times in Columbus). And, in a few years, Jumoke will be ready to give birth again, and Kebi Moyo will have reached breeding age. "With mothers raising babies, or surrogating babies, you're talking four, five, six years between babies," Armstrong says. "So, this could continue another ten years." In any event, even if the surrogate program must be halted until the proper silverback is found, age-diversified social groups will continue to exist in Columbus.

Any consideration of the future of the Columbus gorilla program must also recognize the importance of the zoo's newly expanded great ape facility. The state-of-the-art structure not only enhances the experience for the zoo visitor, allowing humans closer access to the gorillas, but also creates a new environment for the animals and the keepers, opening up new opportunities for the program's expansion and growth.

As it has for nearly half a century, the Columbus Zoo will continue to provide knowledge of and appreciation for gorillas to millions of zoo visitors, both in Columbus and in many other cities with zoos that have been influenced by the Columbus gorilla family. And as it has for the more than forty years since Columbus stunned the zoological community with the announcement that Colo had been born, Columbus's wonderful family of gorillas will continue to fascinate us, teach us important lessons, and earn the respect and admiration of everyone who comes to know them.

COLUMBUS BIRTHS by MOTHER

Mother	Offspring	Birth Date	(Father)
Millie Christina	Colo	12–22–56	(Baron Macombo)
Colo	Emmy	2–1–68	(Bongo)
	Oscar	7–18–69	(Bongo)
	Toni	12–28–71	(Bongo)
Toni	Cora	4–25–79	(Oscar)
	Kahn	7–26–80	(Oscar)
	Zura	9–13–81	(Oscar)
	J.J.	1–2–87	(Sunshine)
	Norman	6–23–88	(Sunshine)
	Jumoke	11–10–89	(Sunshine)
	Nkosi	9–26–91	(Sunshine)
Bridgette	Mosuba	10–26–83	(Oscar)
	Macombo II	10–26–83	(Oscar)
	Motuba	1–25–85	(Oscar)
	Fossey	8–13–86	(Bongo)
Joansie	Roscoe	7–31–80	(Oscar)
	O.J.	7–31–81	(Oscar)
	Lang	12–31–82	(Oscar)
Pongi	Mwelu	7–16–86	(Oscar)
	Colbridge	10–8–87	(Oscar)
	Casode	8–16–93	(Oscar)
Lulu	Lusi	2–27–87	(Sunshine)
	Binti Jua	3–17–88	(Sunshine)
	Kebi Moyo	1–31–91	(Sunshine)
Molly	Unnamed	11–15–92	(Sunshine)
	Unnamed	10–16–93	(Sunshine)
Jumoke	Jontu	1–27–97	(Annaka)

COLUMBUS BIRTHS by FATHER

Father	Offspring	Birth Date	(Mother)
Baron Macombo	Colo	12–22–56	(Millie Christina)
Bongo	Emmy	2–1–68	(Colo)
	Oscar	7–18–69	(Colo)
	Toni	12–28–71	(Colo)
	Fossey	8–13–86	(Bridgette)
Oscar	Cora	4–25–79	(Toni)
	Kahn	7–26–80	(Toni)
	Roscoe	7–31–80	(Joansie)
	O.J.	7–31–81	(Joansie)
	Zura	9–13–81	(Toni)
	Lang	12–31–82	(Joansie)
	Mosuba	10–26–83	(Bridgette)
	Macombo II	10–26–83	(Bridgette)
	Motuba	1–25–85	(Bridgette)
	Mwelu	7–16–86	(Pongi)
	Colebridge	10–8–87	(Pongi)
	Casode	8–16–93	(Pongi)
Sunshine	J.J.	1–2–87	(Toni)
	Lusi	2–27–87	(Lulu)
	Binti Jua	3–17–88	(Lulu)
	Norman	6–23–88	(Toni)
	Jumoke	11–10–89	(Toni)
	Kebi Moyo	1–31–91	(Lulu)
	Nkosi	9–26–91	(Toni)
	Unnamed	11–15–92	(Molly)
	Unnamed	10–16–93	(Molly)
Annaka	Jontu	1–27–97	(Jumoke)

Index

Numbers in italics refer to pages with photographs.